Having Fun with
MAPS AND GLOBES

A Handbook of Strategies and Activities for Teaching Map and Globe Skills

by Abraham Resnick

Consultant: Robert LaRue,
Social Science
Education Consortium, Inc.

toExcel
San Jose New York Lincoln Shanghai

Dedicated to all who are fascinated by maps and globes and who recognize their importance in understanding the interrelationships of the earth's cultural and physical features.

Those who have purchased *Having Fun with Maps and Globes* have the publisher's permission to duplicate pages for distribution in the classroom.

Having Fun with Maps and Globes

Published by toExcel Press
an imprint of iUniverse.com, Inc.

For information address:
iUniverse.com, Inc.
620 North 48th Street
Suite 201
Lincoln, NE 68504-3467
www.iuniverse.com

ISBN: 0-595-00280-3

Printed in the United States of America

PREFACE

In response to well-publicized concerns about young Americans' lack of geographic knowledge, the U.S. Congress in 1987 designated November 15 to 21 as Geography Awareness Week. In a resolution unanimously passed by both houses of Congress and signed by then-President Ronald Reagan, national leaders asked educators to focus on "the integral role that knowledge of world geography plays in preparing citizens of the United States for the future of an increasingly interdependent and interconnected world."

Basically, knowledge of world geography includes not only familiarity with where human-made and natural features are located, but also with why they are there, how they affect each other, and how they influence and are influenced by humans. Because maps and globes are the most efficient means of conveying information about location and interaction among locations, knowledge of how to use maps and globes is essential in building a strong geography education program.

Abraham Resnick
Professor of Social Studies Education
Jersey City State College

INTRODUCTION: **Organization and Use of This Book**

The primary purpose of this handbook is to provide you, as educators, with a variety of proven activities to make learning map and globe skills both enjoyable and meaningful for your students. The activities, which include cross-curriculum tasks, are comprehensive, challenging, and versatile. Easily reproducible, they can be used with multiple grade levels, small groups, or individually. In addition, the flexibility of the three-ring binder allows you to add activities and lesson plans that you have developed, thus making the handbook even more useful.

Having Fun with Maps and Globes is organized into six parts. *Part One* introduces you to the basic goals of a Map and Globe Skills Program and concludes with some general "Tips for Teachers."

Part Two provides activities for teaching the basic concepts of maps and globes. Because the ability to read and make maps involves many individual skills, the activities in this section are organized into several categories or subsets of skills. For each subset, several fully developed activities are presented, along with a "grab bag" of additional activities that can be used in a stand-alone map and globe unit or to reinforce general map skills.

Because map and globe skills support other curriculum areas besides social studies, *Part Three* provides activities organized by subject area. This allows you to locate the subject you are teaching and access several related map and globe skills activities. Such cross-curriculum tasks will reinforce art, math, reading, science, language, and thinking skills.

For easy access, all student activity sheets (called Supplements in this manual) are located in *Part Four* of the handbook.

Part Five lists currently available resources for teachers and students. Beyond the standard lists of print materials, the author has provided sources for multimedia kits and computer software that will enrich anyone's program.

An appendix, which includes map masters, a glossary of terms, and comparison charts, concludes the handbook. Although these materials are mentioned in various activities throughout the book, here they are more accessible as a reference and as a source from which to draw in developing your own lessons, or an entire school program.

Because of the wide range of resources *Having Fun with Maps and Globes* provides, this handbook is a wonderful companion to the supplementary materials you presently use in your school curriculum.

NOTE: This handbook is designed as an ongoing teacher resource: it is not meant to be used as a complete reference for teaching map and globe skills. Additional sections, which will improve, extend, and offer new approaches to the existing information, are currently being written.

Part ONE

Goals of a Map and Globe Skills Program

"In a democracy the development of compassionate and effective public policies depends upon active participation of citizens who are broadly educated about their own society and its relations with the entire world. All events affecting society occur within a geographic context. To understand these events fully we must subject them to geographic scrutiny. . . . Reading, interpreting, and making maps are skills integral to geographic education and to acquiring geographic knowledge."

Guidelines for Geographic Education
(Association of American Geographers and National Council for Geographic Education, 1984)

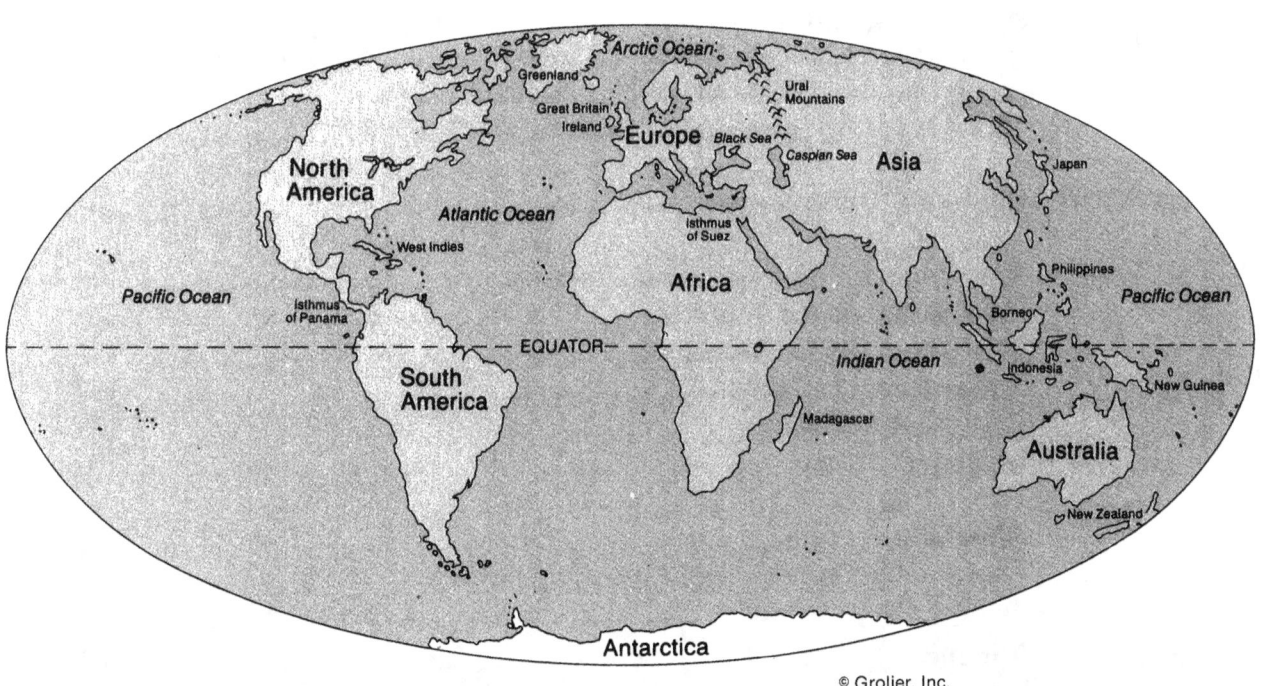

CONTENTS

Preface 5
Introduction 6

Part ONE   **Goals of a Map and Globe Skills Program** 7
             Tips for Teachers 11

Part TWO   **Teaching the Basic Concepts** 13
             Working with Symbols and Legends 13
             Understanding Direction and Location 19
             Working with Scale and Distance 26
             Using Grids 30
             Interpreting Maps 37
             Making Maps 43

Part THREE   **Cross-Curriculum Activities** 51
             Using Maps at Home, at School, and in My Neighborhood 51
             Using Maps to Learn about Our Community 63
             Using Maps to Learn about Our State 67
             Using Maps to Learn about U.S. and World Geography 73
             Using Maps to Learn about U.S. History 80
             Using Maps to Learn about Current World Events 86

Part FOUR   **Student Supplements** 91

Part FIVE   **Resources for Teachers and Students** 165
             Print Materials for Teachers 165
             Print Materials for Students 167
             Audiovisual Materials 168
             Computer Software 169
             Multimedia Kits 170

Part SIX   **Appendix** 173
             Fascinating Facts about Maps and Globes 173
             Making Comparisons 175
             Map Masters 181
             Glossary 193
             Index 198

Part ONE

Goals of a Map and Globe Skills Program

The primary goal of teaching map and globe skills is to help students acquire and interpret information from maps that will be useful in their daily lives, whether that involves understanding the world around them or solving simple problems of distance and location. While few students will become cartographers, most will need to understand and communicate spatial information in their daily lives. Therefore, knowledge of how to use maps and globes is essential.

The following lists identify the basic geography skills and concepts that students should learn in grades K–8. It should be kept in mind, however, that they are intended as guides, not prescriptions, for planning and evaluating instruction.

Skills in the intermediate and middle-level lists depend on mastery of the items in the preceding list. Thus, the K–3 list might be used to assess what remediation is needed before students begin learning the concepts prescribed for the intermediate grades; both the K–3 and 4–6 lists can be used for this purpose at the junior high level.

Basic Concepts for Primary Grades (K–3)

Students in the primary grades should be able to:

1. Understand that the earth is a sphere.

2. Explain that the earth's rotation causes day and night and that its revolution around the sun creates changes in seasons.

3. Recognize that the globe is a model of the earth.

4. Understand that large areas of the earth are represented by smaller areas on a map.

5. Recognize that scale indications on a map show how real distances have been changed into map distances.

6. Use relational terms (e.g., *near, far, long, short, large, small*) to describe distances and areas shown on maps.

7. Understand that map symbols represent real objects on earth.

8. Identify different kinds of symbols used on maps such as colors, lines, shapes, pictures, and points.

9. Use a map key to identify what symbols represent.

10. Use color symbols to distinguish between land and water areas on a map or globe.

11. Distinguish between natural and human-made features on a map.

12. Name the cardinal and intermediate directions.

13. Use the sun or a compass to locate the cardinal and intermediate directions.

14. Given north, locate the other directions within the classroom, school, or neighborhood.

15. Use a compass rose to identify directions on a map.

16. Use cardinal and intermediate directions to identify locations in the environment or on a map.

17. Locate landmarks and important features on a map of the local community.

18. Identify the states and some other features on a U.S. map.

19. Identify the continents and the United States on a world map.

20. Understand that grids are used to locate places on a map.

21. Use a simple alphanumeric grid to find places on a map.

22. Create symbols.

23. Draw a simple map of the home, school, or neighborhood.

Basic Concepts for Intermediate Grades (4–6)

Students in the intermediate grades should be able to:

1. Identify the direction of the earth's rotation as eastward and explain the relationship of rotation to day and night, sunrise and sunset.

2. Explain how the earth's inclination and revolution cause seasonal changes.

3. Explain why a globe is the most accurate way of depicting the earth.

4. Define a map as a scale drawing that uses symbols to show where places are located.

5. Compare the advantages and disadvantages of maps and globes: Explain why distortion is a disadvantage.

6. Recognize that the shape of mapped objects stays the same regardless of the map's scale.

7. Use a bar scale and ruler to estimate distances on a map.

8. Use scale expressed as a ratio to estimate distances on a map.

9. Locate the same place on maps of different scales.

10. Explain the use of map insets.

11. Use the map title to identify the purpose of a map.

12. Use the map legend to identify the symbols used on a map.

13. Interpret a variety of map symbols.

14. Recognize that type and dot size indicate the relative size of cities.

15. Find and use standard symbols.

16. Distinguish between political and physical features on a map.

17. Distinguish between human-drawn and natural boundaries.

18. Use common map abbreviations.

19. Given north, identify the intermediate directions.

20. Use the compass rose or a direction arrow to identify the cardinal and intermediate directions on a map.

21. Use the cardinal and intermediate directions to locate places on a map.

22. Identify the direction traveled between two locations on a map.

23. Use directional terms to trace routes.

24. Locate specific natural or human-made features on a state map, U.S. map, and world map.

25. Compare the geographic features from one section of a map to those shown in an aerial photograph of the same area.

26. Use an alphanumeric grid to locate places on a map.

27. Use the map index to locate places on a map.

28. Show how latitude and longitude create a global grid system.

29. Show how latitude is designated as degrees north or south of the equator and how longitude is designated as degrees east or west of the prime meridian.

30. Identify the equator, prime meridian, poles, tropics, and Arctic and Antarctic circles on a map and on a globe.

31. Show how the earth is divided into hemispheres, and identify the hemispheres.

32. Use latitude and longitude to locate places on a map.

33. Tell why lines of latitude and longitude curve on a flat map.

34. Given a specific purpose, select a map or globe for use.

35. Recognize that there are many different kinds of maps and distinguish among them.

36. Gather data from such special-purpose maps as weather maps, elevation maps, precipitation maps, population maps, economic activities maps, and so on.

37. Use maps to describe the physical characteristics of an area and draw inferences about life in the area.

38. Use an atlas.

39. Recognize the relationship between elevation and the flow of rivers and between elevation and climate.

40. Use a map to plan a trip.

41. Compare maps and draw inferences.

42. Describe different kinds of maps and their uses.

43. Draw maps that include titles, scales, legends, and direction indicators.

Basic Concepts for Junior High Grades (7–8)

Students in the junior high grades should be able to:

1. Describe the earth's angle of inclination and its effects.

2. Draw the earth's orbit and explain its effects.

3. Define projection and understand that every projection distorts at least one of the characteristics shown on a map—shape, size, distance, and direction.

4. Compare projections and identify which characteristics each distorts.

5. Explain why distortion is greater on maps of larger areas than on maps of smaller areas.

6. Choose the best projection for a particular purpose.

7. Explain why the scale used on a map may vary from the equator to the poles.

8. List the advantages and disadvantages of large- and small-scale maps.

9. Understand and use contour lines.

10. Recognize that the compass rose can be divided into additional subdivisions (ESE, ENE, etc.).

11. Identify directions on a polar projection map.

12. Compare geographic features found in satellite images with those found on maps of the same region.

13. Identify great circle routes on a globe and a map.

14. Relate longitude to time zones and locate the International Date Line on a map.

15. Use the time dial on the globe and the time at a reference point to find the time at other selected points.

16. Use a time zone map to compute time changes during international travel.

17. Use latitude to calculate distances.

18. Use special-purpose maps such as underwater landforms, language/cultural regions, time zones, ocean currents, trade winds, and so on.

19. Use maps to infer how climate is affected by ocean currents, prevailing winds, and insolation belts.

20. Use data from maps to explain how landforms, climate, and historical development influence one another (e.g., identify barriers to settlement, factors affecting location of major cities).

21. Infer from a description of an area where it might be located on a map.

22. Use specialized atlases.

23. Make predictions using information from maps.

24. Participate in field mapping, pacing, and orienteering activities.

25. In preparing maps, use simple map conventions (e.g., placing names of cities to the right of their symbols, placing names of rivers parallel to their courses).

26. Consider design factors (e.g., the fewer the colors, the clearer the meaning) in mapping.

Tips for Teachers

The following ideas may be helpful in creating a geography-rich environment in your classroom:

1. Start a map skills learning station (a cartographer's corner) containing miscellaneous media, instruments, and reference aids. Items that might typically be kept in such a learning center are listed at the end of this section.

2. Place direction placards on appropriate classroom walls to assist students in orienting maps they are using. Although students should not be encouraged to think that north is at the top of every map, hanging maps on the north wall may help them to retain a correct east-west orientation.

3. When new maps and globes are first displayed in the classroom, give students the opportunity to analyze, interpret, and discuss them. Encourage them to draw inferences from the maps and use them as needed in problem-solving exercises.

4. Relate map work to real-life situations and events. For example, use maps in conjunction with field trips. Prior to a trip have students draw the route they will take. If a map is available of the site they will be visiting, study it. Allow students to carry copies of the map with them during the visit. When the students return to the classroom, ask them to refer to the map when they are sharing their experiences.

5. Use maps to support other subject areas of the curriculum. In language arts, maps can be used to locate the setting of a story. They are extremely useful in science classes, when studying such topics as ecology, weather, and natural resources. In art classes, drawing maps will help students learn concepts such as scale, perspective, and color.

Items for a Cartographer's Corner

A typical collection of items that may be kept in class kits or in cartons at the cartographer's corner learning center might include:

globes	acetate transparency	map skill activity books
art boards	sheets and rolls	magnifying glasses
global measuring	opaque projector	symbol charts
templates	overhead projector	large wall-size news maps
protractors	folding maps	current events magazines
rulers	map file	pedometers
No. 2 pencils	gazetteers	kraft paper
colored marking pens	U.S Geological Survey	map puzzles
triangles	Topographic Sheets	map and geography
graph paper	charts	games
road maps	dictionary	computer software
desk outline maps	geographical terms chart	map scale measuring dial
small, handled knife for	compasses	lettering pens
cutting paper	binoculars	stencils
scissors	six-foot tape measure	geographic encyclopedias
pictures	carbon paper	atlases
tagboard sheets	tissue paper	National Geographic
construction paper	glue and paste	magazines
string	acetate sheets	stapler

Part TWO

Teaching the Basic Concepts

Part TWO

Teaching the Basic Concepts

The activities presented in this chapter are not intended as material for a complete map and globe skills program. Rather, their purpose is to provide you with options for teaching, reinforcing, or remediating basic concepts in map and globe skills. These activities will also spark new ideas.

The activities are organized into the following sections:

- Working with Symbols and Legends
- Understanding Direction and Location
- Working with Scale and Distance
- Using Grids
- Interpreting Maps
- Making Maps

Each section contains several complete lessons, followed by numerous suggested activities. Some encompass more than one skill area; each is classified according to the skill it treats most extensively.

Working with Symbols and Legends

In teaching students about symbols and legends, you will find it helpful to have a collection of materials that use symbols such as maps, magazine or newspaper advertisements, and pictograms. You can begin such a collection yourself and then have students add to it throughout the year.

Activity 1-1-2 *The globe: model of the earth*

Purpose: To understand the concept of *model* and how a globe is a model of the earth

Materials:

1. one or more globes
2. satellite picture of the earth
3. several examples of other kinds of models: cars, airplanes, stuffed animals, instructional models

Procedure: Ask students to identify the various objects you have on display.

When they have done so, ask if they can think of one word that would describe all of these objects (models). Define the word *model* (an object that is a smaller or larger representation of another object). Ask students why we use models. (The real object may be too large, too small, too expensive, or too dangerous to study or play with. Some models emphasize things that individual objects have in common.) Have students identify what each of the models represents.

Next, display the globe and the satellite photo of the earth. Ask students how the two are related. (The globe is a model of the earth.) On the globe have students find the portions of the earth that are shown in the photograph. Then ask: In what ways is the globe like the earth? (It is round, it has an axis, it spins on its axis.) In what ways is the globe different from the earth? (It is much smaller; it is made of only a few substances instead of many different substances.)

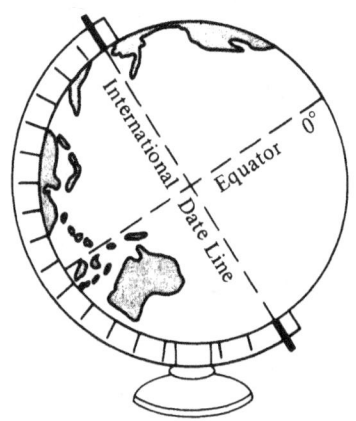

Discuss why the globe is a useful tool. It shows where places are located, relative sizes of places, relative direction, and it is a convenient size to study.) Ask students how the globe shows information. (It uses symbols.) Have students identify some of the symbols used on the map and what they represent. Encourage questions. Make sure each student has an opportunity to examine a globe.

Follow-up: You may ask older students to compare the globe with a world map, noting similarities and differences, especially regarding symbols. Younger students might construct a globe. Instructions are given in the Making Maps section at the end of Part Two.

Activity 2-1-2 *Photographs, drawings, and maps*

Purpose: To differentiate among photographs, drawings, and maps of the same location

Materials:

1. an aerial photograph, a drawing, and a map of the same area

If you find a good aerial photograph with a moderate amount of detail, you could prepare the drawing and map yourself.

Procedure: Display the aerial photograph. Ask students to identify how this picture was made. (It was taken with a camera.) Where was the photographer when he/she took this photograph? (somewhere above the ground) What is this kind of photograph called? (an aerial photograph) Use the chalkboard to list the features shown in the photograph.

Next, display the drawing. Go through a similar series of questions and conclude by comparing the features shown in the photograph and drawing. Be sure students understand that the artist can decide whether to include features or leave them out, which is why a photograph and drawing of an area might not show all the same features even if taken/drawn from the same vantage point.

Finally, display the map and go through a similar series of questions. Focus a concluding discussion on the use of symbols on the map and the ability of the cartographer, or mapmaker, to be very selective in deciding what to show on a map.

Follow-up: Provide students with a selection of other aerial photographs (or ask them to find such photographs) and have them draw maps based on the photographs.

Activity 3-1-2 *Symbols and road signs*

Purpose: To understand the various uses of symbols and identify their meaning on road signs

Materials:

1. copies of Supplement 3-1-2
2. paper, crayons, or color markers

Procedure: Ask students to explain why symbols rather than words are used on so many road signs (so that everyone can understand them, whether they can read the language or not).

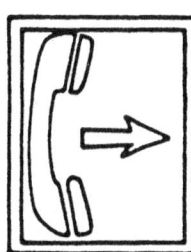 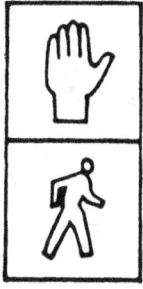 

Divide the class into six groups and distribute Supplement 3-1-2. Assign each of the groups five of the items on the list. Students are to match each symbol with its correct message.

Follow-up: Encourage students to use the library or other sources to find additional symbols not shown on the handout. You might assemble the class road sign symbols into a booklet or chart. Students might also have fun drawing road sign symbols in place of some words as they write safety essays.

Activity 4-1-2 *Using flags as symbols of countries*

Purpose: To identify the political boundaries of countries on a world map and understand that flags are symbols of countries

Materials:

1. world map
2. toothpicks, clay, paper, crayons or color markers, tape
3. reference books
4. index cards with the names of several countries written on each

You may select the names of the countries from the United Nations Membership list, which can be found on Supplement 4-1-2.

Procedure: Direct students' attention to the world map. Discuss political and natural boundary lines. Be sure students recognize both the line symbol and the use of color to designate boundaries.

Ask students to find symbols of our country displayed in the school or classroom (flag, eagle, Uncle Sam). Discuss how the flag uses symbolism: thirteen stripes represent the thirteen original states; stars represent the fifty states; the colors red, white, and blue represent valor, purity, and justice.

Divide the class into groups of three or four students and give each group a set of the index cards. Ask students to do research on the flags of the countries typed on their cards. After having made a list of the symbols used in the flags, students should make miniature flags, using toothpicks, paper, and tape.

When the models are complete, have students locate their countries on the large world map or globe and attach their flags by placing a piece of clay on the country and inserting the toothpick with the flag into the clay. Allow time for each group to report on the symbolism used in their flags.

Follow-up: Students could undertake a similar project for the fifty U.S. states.

Activity 5-1-2 *Using legends and symbols*

Purpose: To locate legends on maps and recognize various kinds of symbols, comparing their uses on several maps

Materials:

1. five or more atlases

Procedure: Review the definition of a symbol (a line, picture, dot, color, or shape that represents something else). Point out that every map has a key, or legend, that explains the symbols that are being used on that particular map. Show examples of legends in an atlas. Supplements 5-1-2(a), (b), (c), (d), (e), will provide many types of map symbols.

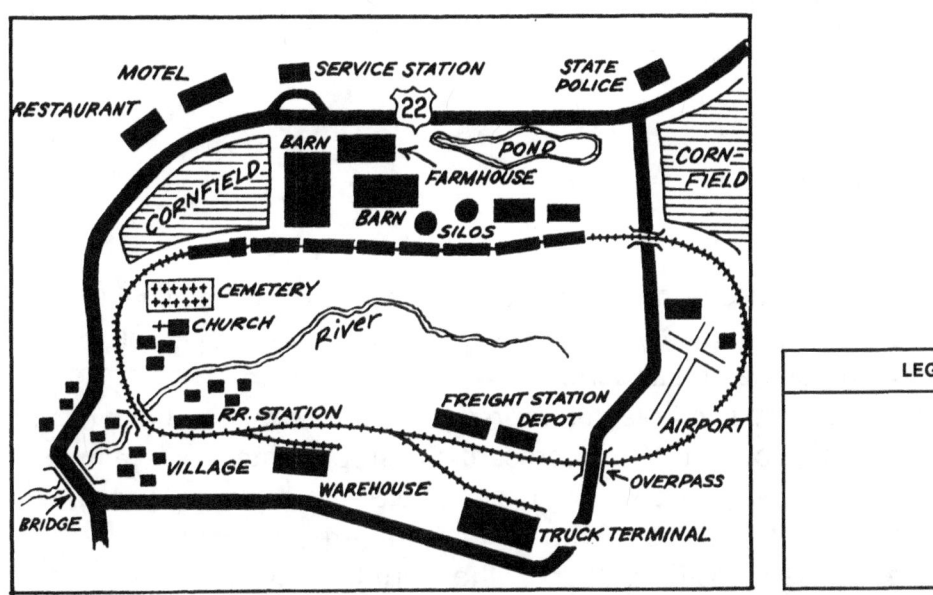

Divide the class into groups and give each group an atlas. Tell them to make a chart, similar to the one below, and fill it in using symbols they find in the atlas.

Kind of Symbol	Examples	What Examples Stand For
line		
color		
dot		
pattern		
shape		

Have students share some of the symbols they found. Then ask them to give examples of how they might use the same symbols to represent different things on different maps. For example, the color green might mean one thing on a political map and another thing on a physical map.

Tell students that there are some symbols that always represent the same thing on any map. They are called standardized symbols. Have students use atlases to find the following standardized symbols:

> water
> capital city
> other city or town
> international boundary

Follow-up: Have students create symbols that can be used on various thematic (special purpose) maps. For each symbol chosen, students should draw two variations of it: one that looks like what it represents and one that does not. For example, for a population map, students might create a small human figure as one symbol and a dot or color as another.

Additional Activities

- *Classroom map*

Make a map of your classroom, complete with desks, windows, chalkboards, and so on. Distribute the maps to the students and discuss the symbols being used. Draw a legend at the chalkboard.

Ask students to find the symbols that represent their desks and those of their friends. Then design a color code to represent classroom data. For example, color all girls' desks purple; boys' desks, green. Color the desks of students who prefer pizza, red; the desks of those who prefer hamburgers, brown. Provide additional copies of the maps for students who want to show other kinds of information.

- *Tomorrowland*

Have students design and draw a future map of their local community, city, county, state, or a mythical place, as projected for the 21st century.

The hypothetical map of tomorrowland might include such features as hydrofoil boat docks, monorails, helicopterports, surface trams, overhead cable-car lines, moving walkways, space stations, two-tier highways, underground shopping centers, etc.

The use of pictorial symbols should be encouraged.

- *Commercial maps and floor plans*

Make or acquire maps of local shopping malls or business districts. Go over the maps with the students, identifying each building or shop and writing its name on the map.

Ask students to get brochures, business cards, or advertisements for each store. Review the materials, pointing out how logos are used as symbols. Have students cut out the logos and attach them to the map in the respective places. Discuss some of the logos. Are they appropriate symbols for the stores they represent? Why/why not?

You can conduct a similar activity using the floor plan for a large store and symbols from the various departments.

- *Scrambled states*

Maps often use abbreviations in order to save space; thus the abbreviations become symbols. To provide practice in using the abbreviations and locating the states, distribute copies of Supplement 6-1-2 and a political map of the United States from the Map Masters section of the appendix. Have students complete the exercise.

Understanding Direction and Location

The activities in this section provide practice in learning directions and using them to locate places on maps and globes. Eventually, students should become skillful in locating places that are in the news and be able to explain their geographic importance.

Activity 1-2-2   *The earth's movement*

Purpose: To understand how the movement of the earth determines seasons and day and night

> **Materials:**

1. a globe
2. a flashlight

Procedure: Display the globe. Explain that the globe is a model of the earth. Point out the North and South poles, explaining that north is toward the North Pole and south is toward the South Pole.

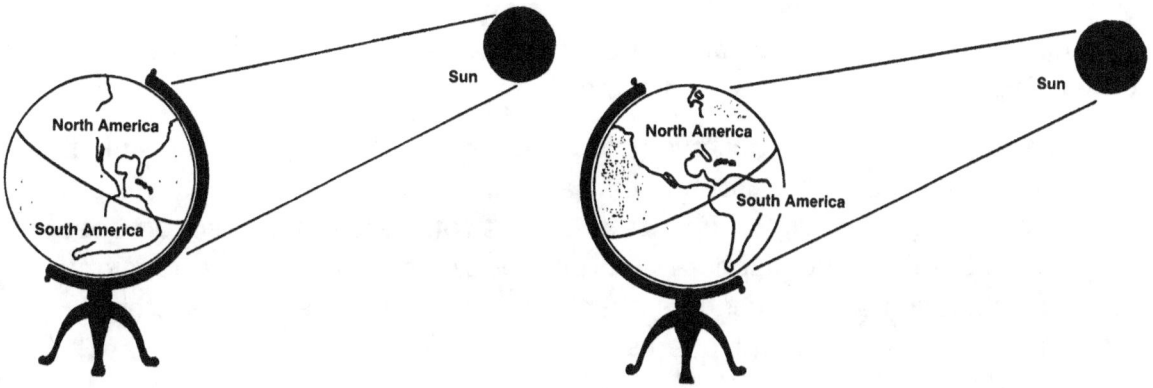

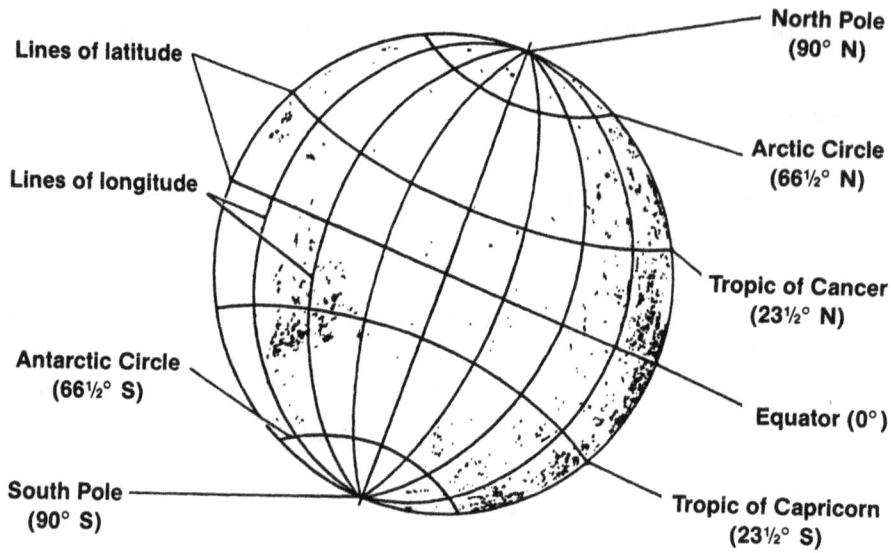

Ask students if there are East and West poles. When they respond that there are no such poles, ask how the directions east and west are determined. Students are likely to respond that east is the direction in which the sun rises and west is the direction in which it sets. Explain that this is true because the earth rotates from west to east. Demonstrate this by spinning the globe in the proper direction.

This idea can be reinforced by having students hold their right hands with the thumb pointed north. When they close their fingers to make a clenched fist, the fingers rotate from west to east, just as the earth does.

Demonstrate the rotation of the earth and its effect on day and night by darkening the room and directing the beam of a flashlight toward the globe. Discuss with students what happens when the earth rotates. When the sun rises in Japan, for example, what is happening in the United States?

Expand the demonstration to include revolution of the earth around the sun. Have two students move the globe around the flashlight (sun); one student should steadily walk the globe in a slightly flattened circle while the other rotates it in the correct direction. As the earth revolves around the sun, what happens at various points in the orbit? What is happening in the areas tilted away from the sun? (winter) What is happening in the areas tilted toward the sun? (summer) With older students, you may wish to go into a more detailed discussion of the equator, tropics, and Arctic and Antarctic circles, developing generalizations about latitude and climate.

Follow-up: Have students observe how water swirls down a drain in a sink at home, and have them check other sinks as well. Why does it always swirl clockwise as it goes down? (because of earth's rotation) What direction does it swirl in the Southern Hemisphere? (counterclockwise) Older students might look for relationships between this phenomenon and wind patterns in the Northern and Southern hemispheres.

Ask students to investigate the symbolism in the Japanese flag. How does it relate to what they learned in this lesson?

Activity 2-2-2   *Using a compass rose*

Purpose: To identify the cardinal and intermediate directions on a compass rose and use them to locate places on a map

Materials:

1. copies of Supplement 2-2-2
2. color markers or crayons
3. a compass
4. direction placards

Before beginning this activity, draw a large compass rose on the chalkboard. Label only north on the compass rose.

Procedure: Using the compass, review with students the cardinal directions, identifying and labeling north, south, east, and west in your classroom.

Draw students' attention to the compass rose on the chalkboard. Ask volunteers to write in the other three cardinal directions. Review or introduce the intermediate directions and have volunteers add them to the compass rose.

Distribute Supplement 2-2-2 and have students complete the top half of the worksheet. Were students surprised that north was not always at the top of the compass rose? Explain that while north is commonly at the top of maps, it does not have to be; students should be sure to orient themselves to any new map by checking the compass rose to see where north and the other directions are located.

Make sure students have blue, green, red, and black crayons or markers. Have them follow these directions to complete the bottom half of the worksheet:

- Label the correct sides of the grid with an E, S, and W.
- Draw a line from the N to the E.
- Go one square west from the E. Then go one square north.
- Label that point NE. Color the square to the southwest black.
- Draw a line from the S to the E.
- Go two squares west from the E. Color the square to the southwest green.
- Draw a line from the S to the W.
- Go one square east of the W. Color the square to the northeast blue.
- Draw a line from the W to the N.

- Go two squares south of the N. Color the square to the southeast red.
- Label the corners of the red, green, and blue squares with the correct abbreviations for the intermediate directions.
- Give this direction-finding tool a name. (Accept all answers.)

Follow-up: Have students use the tools developed in this activity to play a game of "Directional Simple Simon." For example, instructions might include "Simple Simon says point northeast" or "Point toward the direction shown by the blue square."

Activity 3-2-2 *Traveling by directions*

Purpose: To use cardinal and intermediate directions when tracing a route on a map

Materials:

1. large 10' x 10' (or larger) floor map depicting an imaginary setting (city, jungle, farm, zoo). (See sample below.) Map should include a legend and a compass rose.
2. game markers

Conduct this activity in an area that will allow students access to the floor map.

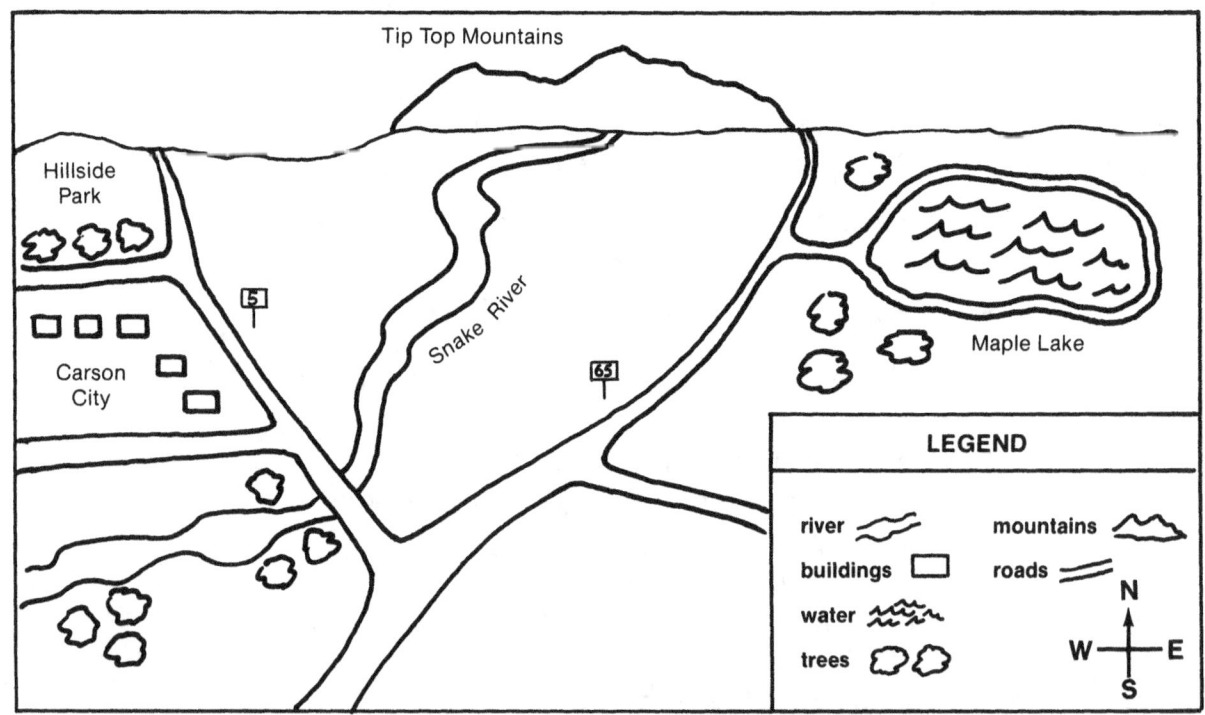

Procedure: Place the map on the floor. Allow students time to explore it. Discuss using the legend and the compass rose.

Ask one student to place a marker from a game board in the middle of the map. Following your directions, the student should move the marker to various locations. (Use cardinal and intermediate directions, as well as references to features shown on the map in giving your directions.) For example, directions might include the following: "Go east until you get to Highway 65. Take I-65 to the mountains. Then go west to the river. Take the river southwest to Highway 5. Finally, proceed northwest into Carson City."

Have students take turns making up directions and following them. Or, have them draw their own maps.

Follow-up: Encourage students to write diary entries describing their travels through the city, zoo, or jungle shown on the floor map. Their entries should be based on map information, as well as their own imaginations.

Activity 4-2-2   *Tracing routes*

Purpose: To acquire skills in tracing routes on a U.S. map

Materials:

1. copies of Supplement 4-2-2
2. laminated U.S. desk maps showing major cities
3. water-soluble marking pens

If laminated desk maps are not available, you may use the political U.S. map in the Map Masters section of the appendix. Each student should have two copies of the map.

Procedure: Discuss the importance of being able to use directions. Tell students that they will be tracing routes on a map of the United States. To do so, they will have to follow directions and locate major cities. Distribute Supplement 4-2-2 and the laminated desk maps or the two copies of the U.S. map. Ask students to complete the activity.

Answer key for Supplement 4-2-2: Part 1: star; Part 2: (G, O, O, D, W, O, R, K).

Follow-up: Invite a person from the community to speak to the class about how he/she uses maps and directions at work or as part of a hobby. The person can be a pilot, bus driver, police officer, astronomer, explorer, travel writer, captain of a ship, and so on.

Additional Activities

- *Literary geography*

Post a world map to which you will refer when locating the settings for stories or biographies that students read. Have the students mark the locations with symbols that represent the people or places in the stories. Students might also draw individual maps of the places featured in the stories; these can be floor plans of the homes of characters or maps of the communities in which the characters lived.

- *Map on the back*

Have each student select a specific place on a state, United States, or world map. Each student should either draw the outline of the state or country, or print its name in large letters on a sheet of paper. The papers should then be stacked face down at the front of the room.

Ask a student to come to the front of the room. After you have taped one of the papers from the stack on his/her back, have the student face the map, exposing the name of the place location to the rest of the class.

The student can ask a maximum of fifteen questions to help him/her identify the place written on the paper attached to his/her back. Questions can only be answered with *yes* or *no*. Some of the questions students might ask are as follows: Is it a continent, country, state? Is it a physical feature? Is it east of. . . ?

You could divide the class into teams and give them one point for every question asked of each player. The team with the fewest points wins.

- *Hidden words*

Distribute Supplement 5-2-2. Ask students to use a world map or atlas to help them complete the activity. They are to read the clues for each word and then rearrange the letters of the words to create names of states or countries. Students might like to work together in pairs. Once they have completed the activity, ask them to create words of their own.

Answer key for Supplement 5-2-2: 1. Paris, 2. Peru, 3. Texas, 4. Spain, 5. Reno, 6. Manila, 7. Burma, 8. Lima, 9. Rome, 10. Oslo, 11. Laos, 12. Israel, 13. Niger.

- *Football*

Draw a large football gridiron on the chalkboard. Divide the class into two-, four-, or six-member teams that will be competing against each other to score points. Before each set of two teams faces off, write their team names at each end of the gridiron. Tell the team members to stand near a large wall map of the world or the United States while you or other students call out a place location; the first team to find it on the map advances ten yards on the gridiron. Winners must advance to the competing team's goal line.

- *Treasure hunt map game*

Divide the class into groups of three or four students, giving each a treasure to hide somewhere in the classroom or playground. Group members should decide where they will hide their treasure and then draw a pirate's map showing the route to take to find the treasure. (The maps should show fixed points in the classroom or playground, the number of steps from point to point, directions, and so on.)

Let groups secretly hide their treasures and then exchange maps with other groups. Allow students time to search out the treasures. Keep track of the time it takes each group to find the treasure. Discuss why some students found the treasure faster than others.

- *Where in the world?*

Select thirty or forty diverse places shown on your classroom globe. Write the name of each place on an index card and place the cards in a box. Have students take turns picking cards at random from the box and finding the places on the globe. When a student has found the place listed on the card, he/she should make a geographic statement about the place. Examples might include: "Asia is the largest continent," "Great Britain is surrounded by water," or "The mouth of the Amazon is at the equator."

- *Map searches*

Use Supplements 6-2-2; 7-2-2; 8-2-2 in Part Four of your handbook to provide students with practice in locating places on U.S. and world maps. Answer keys follow each Supplement.

Working with Scale and Distance

Use of a scale requires math ability as well as an understanding of the underlying concept of scale. Once students learn how to use a scale, they should practice frequently. The following activities will provide you with ideas for making that practice possible.

Activity 1-3-2 *Understanding scale*

Purpose: To understand the concept of scale and learn how to use a scale

Materials:

1. copies of Supplement 1-3-2
2. rulers

Procedure: Ask students if they have heard of scale models or maps drawn to scale. Explain that "to scale" refers to the way something large is made into a smaller size either in a drawing or a model.

When something is made to scale, it means that all the parts have been reduced the same amount (proportionately). To illustrate this concept, draw the following two objects on the chalkboard. Ask students to tell you which object has been drawn to scale.

Then draw the two diagrams below on the chalkboard. Tell students towns B and C are the same distance from town A. Ask them which diagram is drawn to scale? Which diagram has been drawn proportionately?

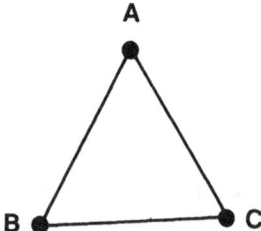 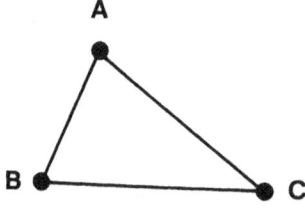

Clarify by explaining that a certain distance on a map drawn to scale represents an actual distance on earth.

Distribute Supplement 1-3-2 and read it with students. Ask them to complete the exercise. Their drawings should be 1" x 1".

Follow-up: Give students graph paper and let them experiment making scale drawings of various shapes.

Activity 2-3-2 *Showing scale on a map*

Purpose: To learn several different ways of showing scale

Materials:

1. copies of Supplement 2-3-2

Procedure: Review the definition of scale. Ask students in what different ways they have seen scale shown? Write all their answers on the chalkboard. Distribute Supplement 2-3-2 and ask students to complete the three exercises.

Answer key for Supplement 2-3-2: 1. 1 inch = 100 miles; 2. 0   25   50 miles; 3. 1:600.

Follow-up: Have students use textbooks and atlases to find examples of the various ways scale can be shown on maps. Challenge the students to convert each scale to another form.

Activity 3-3-2 *Calculating distances on a map*

Purpose: To use scale to calculate distances on a map

Materials:

1. copies of Supplement 3-3-2
2. rulers

Procedure: Review the work you have done thus far with scale. On the chalkboard draw a bar scale like the one above. Explain that this scale shows that 1 inch on the map represents 50 miles on earth. Ask the students how they would calculate distances using the scale? (They would multiply the number of inches between the two points on the map by 50.)

Then draw a series of numbered points on the chalkboard. Ask volunteers to come up and, using the scale you have drawn, measure the distances between the points. Distribute Supplement 3-3-2 and ask students to complete the exercise.

Answer key for Supplement 3-3-2: 1. 15 miles; 2. 25 miles; 3. Louisville and Maple Grove; 4. through Georgetown; about 12 miles; 5. 160 miles.

Follow-up: Suggest additional practice exercises using maps available in your classroom, such as calculating the distance between your community and others in your state or between your state capital and capitals of neighboring states.

Activity 4-3-2   *Using inset maps*

Purpose: To learn the difference between a regular map and an inset map and understand why each map uses a different scale

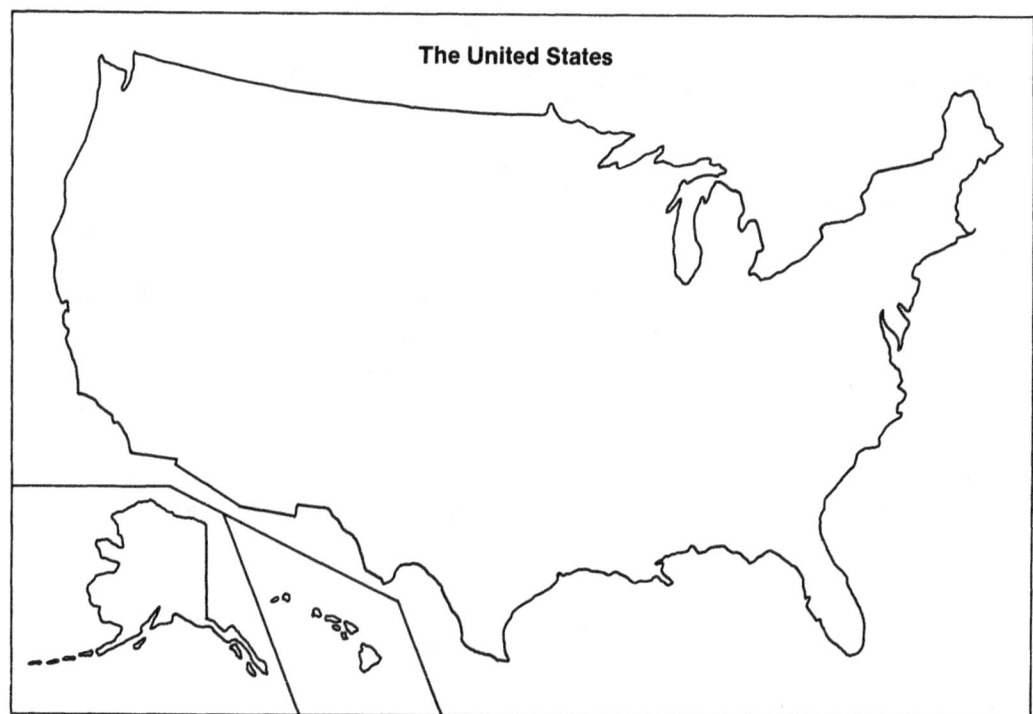

The United States

Materials:

1. wall map of the United States showing Alaska as an inset map
2. road map with city insets
3. one copy of each continent map from the Map Masters section in the appendix
4. atlases and other references

Procedure: Have students look at the wall map of the United States. Ask them to identify the actual location of Alaska relative to the rest of the United States. Why isn't Alaska shown in its actual location? (because the map would become too large or the rest of the United States would have to be made much smaller)

   Tell students that this kind of map is called an *inset map*. Ask a volunteer to come to the map and compare the scales of the inset map of Alaska and the map of the continental United States. Ask students why the map of Alaska has a different scale. (to make it small enough to fit comfortably on the map) How big would Alaska be if it were at the same scale as the rest of the United States? (It would be as large as Texas, Oklahoma, New Mexico, Colorado, and about one-third of Kansas combined.)

Display a road map and ask students to suggest the second reason for using inset maps. (to show more detail about one area)

Divide the class into six groups and give each group a map of one continent. Groups are to use atlases and other reference materials to draw one or more inset maps that show details of a specific location on the continent. The insets should be keyed to the continent maps.

Follow-up: Ask students to prepare posters describing the areas they have chosen to show on the inset maps. The posters might include some descriptive narrative, as well as pictures of well-known features in the area and the map itself.

Additional Activities

- *Making inset maps*

Have students work in pairs to create their own inset maps of the classroom. Have them draw two inset maps: one showing details of a specific feature in the room and the other showing even more detail of a small portion of that feature. For example, one inset might show details in the section of the room where two students sit. The other inset could show details of the top of one of the two students' desks.

- *Planning a trip*

Acquire road maps of several states. Have each student plan a route that incorporates detours, stopovers, side trips, visits, delays, returns, and so on. Instructions for following the route should be written out step by step and include using both directions and scale. Instructions might include such statements as: "Proceed fifty miles on Route 7. Camp overnight at the nearest state park. Then take a side trip to visit the state capital. Return to Highway 28 and drive south 180 miles. Stay with relatives for three days at Newtown."

After students have completed their instructions, they should exchange them with other classmates. Using a toy car or other small object, they should follow the instructions, calculating how long the trip will take. Afterwards, the student who planned the trip should comment on how well the driver followed the map instructions.

- *Calculating distances on a globe*

Point out that measuring distances on a globe requires use of something that can curve, such as a cloth measuring tape or string. Equip students with string, rulers, and globes and have them calculate distances between pairs of national capitals (for example, Tokyo to London, Washington to Brasilia, Moscow to Lagos, Canberra to Buenos Aires).

Using Grids

Grids are important tools for locating places. Two kinds of grids are used with maps: alphanumeric grids, which make use of letters and numbers to locate places, and latitude and longitude grids, which make use of degrees. Because many students have difficulty using latitude and longitude you might have students work with alphanumeric grids first before introducing them to the global grid system. The activities in this section cover both kinds of grids.

Activity 1-4-2   *What is a grid?*

Purpose: To learn how to use an alphanumeric grid

Materials:

1. copies of Supplement 1-4-2
2. crayons, color markers

Procedure: Explain that map grids are tools that help us find places on maps. Draw two sets of intersecting lines on the chalkboard. Explain that a grid is made up of two sets of lines (one set is vertical and the other is horizontal). The lines, which form squares, are given letter and number names.

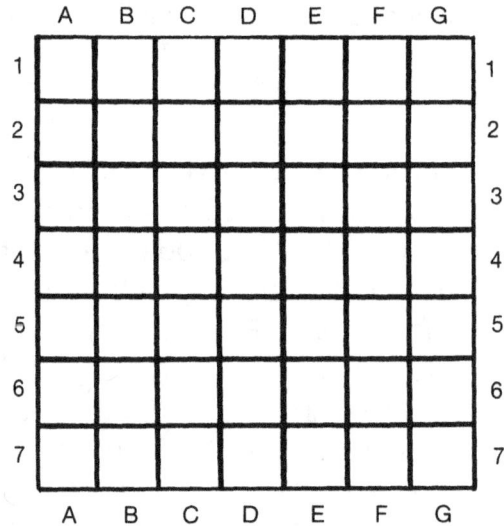

Distribute copies of Supplement 1-4-2 and point out the letters and numbers above and to the sides of the squares. Show how the letters and numbers can be used to name the squares. Explain that, in naming the grid square, the letter is read first and then the number.

Ask students to use their crayons to color the grid according to these directions:

- Color square E-1 blue.
- Color square D-2 blue.

- Color square C-3 blue.
- Color square B-4 blue.
- Color squares A-3, A-4, and A-5 red.
- Color squares B-5 and C-5 red.
- Write your first name in square A-1.
- Write your last name in square E-5.
- Color square B-2 your favorite color.
- Make square D-4 match square B-2.

Answer key follows Supplement 1-4-2 in Part Four of this manual.

Follow-up: Give students additional copies of Supplement 1-4-2 and tell them to make up a series of directions similar to the practice sample above. When they have completed their directions, have them exchange their papers so that other students can complete their grids.

Activity 2-4-2 *Using a grid*

Purpose: To construct a grid on a U.S. map and use it to describe locations

Materials:

1. copies of the political map of the United States from the Map Masters section of the appendix
2. rulers

Procedure: Explain that a grid is composed of two sets of lines—vertical and horizontal—and that grid squares are named using letter-number combinations.

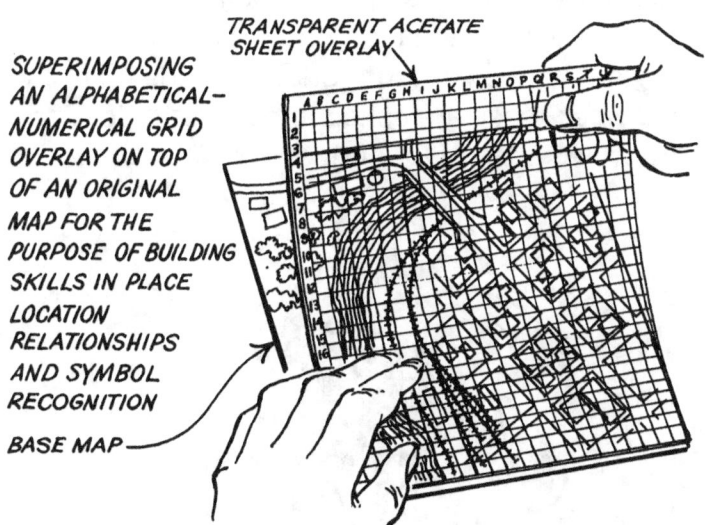

SUPERIMPOSING AN ALPHABETICAL—NUMERICAL GRID OVERLAY ON TOP OF AN ORIGINAL MAP FOR THE PURPOSE OF BUILDING SKILLS IN PLACE LOCATION RELATIONSHIPS AND SYMBOL RECOGNITION

BASE MAP

TRANSPARENT ACETATE SHEET OVERLAY

Distribute the U.S. maps and direct students to draw a grid on the map. They can do so by following these steps:

- Make a mark every inch along the top of the map. Make a mark every inch along the bottom of the map. Connect these marks. Label each column with a letter, beginning with A on the left.

- Make a mark every inch along the right side of the paper. Make a mark every inch along the left side of the paper. Connect these marks. Label each row with a number, beginning with 1 on the top.

While students are creating these grids, write the names of about ten grid squares on the chalkboard. Once a student has completed his/her grid, he/she should locate one state that is completely or partially within each grid square you have listed on the chalkboard.

Follow-up: Have students look at the indexes in atlases or accompanying road maps. Then have them add cities to their U.S. maps and create indexes to accompany the maps.

Activity 3-4-2 *The global grid*

Purpose: To understand that latitude lines, which run east and west, and longitude lines, which run north and south, create a global grid system

Materials:

1. several globes

Procedure: Review the procedure of using an alphanumeric grid. Then introduce the students to a global grid. Explain that, unlike an alphanumeric grid, which uses squares to locate places, a global grid uses points on a globe at which latitude and longitude lines intersect.

Point out that a place is always at the same grid point on a global grid; it does not change from map to map, like its location on an alphanumeric grid can.

Tell students that the global grid, like other grids, is made up of two sets of lines. One goes east and west; these are called lines of latitude. The others go north and south and are called lines of longitude.

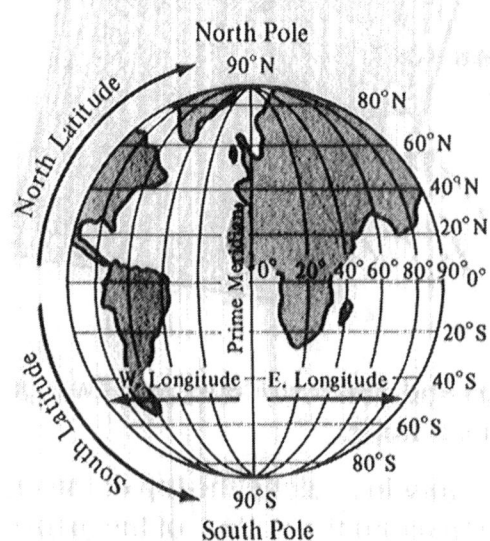

Have students look for lines going east and west on the globe. Point out the equator, which is an imaginary line around the middle of the earth, halfway between the North and South poles. Ask the students to locate 0° on the equator.

Explain that all lines of latitude are numbered from the equator. Between the equator and the North Pole are 90 degrees of latitude. They are numbered from 1 north latitude to 90 north latitude. Have students locate some of these lines of north latitude on the globe. Point out that the lines of latitude in the Southern Hemisphere are numbered in the same way but are labeled south, rather than north.

Next, help students locate the lines going north and south on the globe. These lines are half circles that go from pole to pole. They are numbered from a line called the prime meridian. Ask students to identify what continents the prime meridian passes through.

Point out that lines of longitude are numbered east and west of the prime meridian, up to 180°. Help students locate the 180° longitude line on the globe, as well as some east and west longitudes.

Explain that grid points are the places at which lines of latitude and longitude intersect. They are identified by the number of degrees of the intersecting latitude and longitude lines. To reinforce this concept, have students locate several grid points on the globe and give their names.

Follow-up: Ask students to form groups and work together to make a list of grid points. They should then exchange lists with the other groups and identify on which continent or in which ocean each grid point on the list is found.

Activity 4-4-2   *Using the global grid*

Purpose: To learn how to use latitude/longitude coordinates to locate places on a world map

Materials:

1. detailed world maps showing latitude/longitude lines
2. copies of Supplement 4-4-2
3. detailed U.S. maps, showing latitude/longitude lines
4. copies of Supplement 5-4-2
5. almanac or other reference materials

Procedure: Distribute Supplement 4-4-2 and world maps or Supplement 5-4-2 and U.S. maps. Ask students to complete it.

Follow-up: Encourage students to create exercises similar to Supplement 4-4-2 and have their classmates complete them.

Activity 5-4-2   *Longitude and global time*

Purpose: To understand the relationship between longitude and time zones

Materials:

1. world time-zone map from an encyclopedia or atlas
2. paper, pencils

Before beginning this activity, use an opaque or overhead projector to enlarge the outline map of the world located in the Map Masters section of the appendix. Trace the enlarged map onto butcher paper or newsprint. Using a world time-zone map from an atlas or an encyclopedia, indicate the 24 time zones on the enlarged map.

Procedure: Tell students that the international system of time zones is based on longitude. Explain that there are 24 time zones. Ask students to estimate how many degrees of longitude are in each time zone. To arrive at the correct answer of 15, students may need to be reminded that there are 360 degrees in a circle.

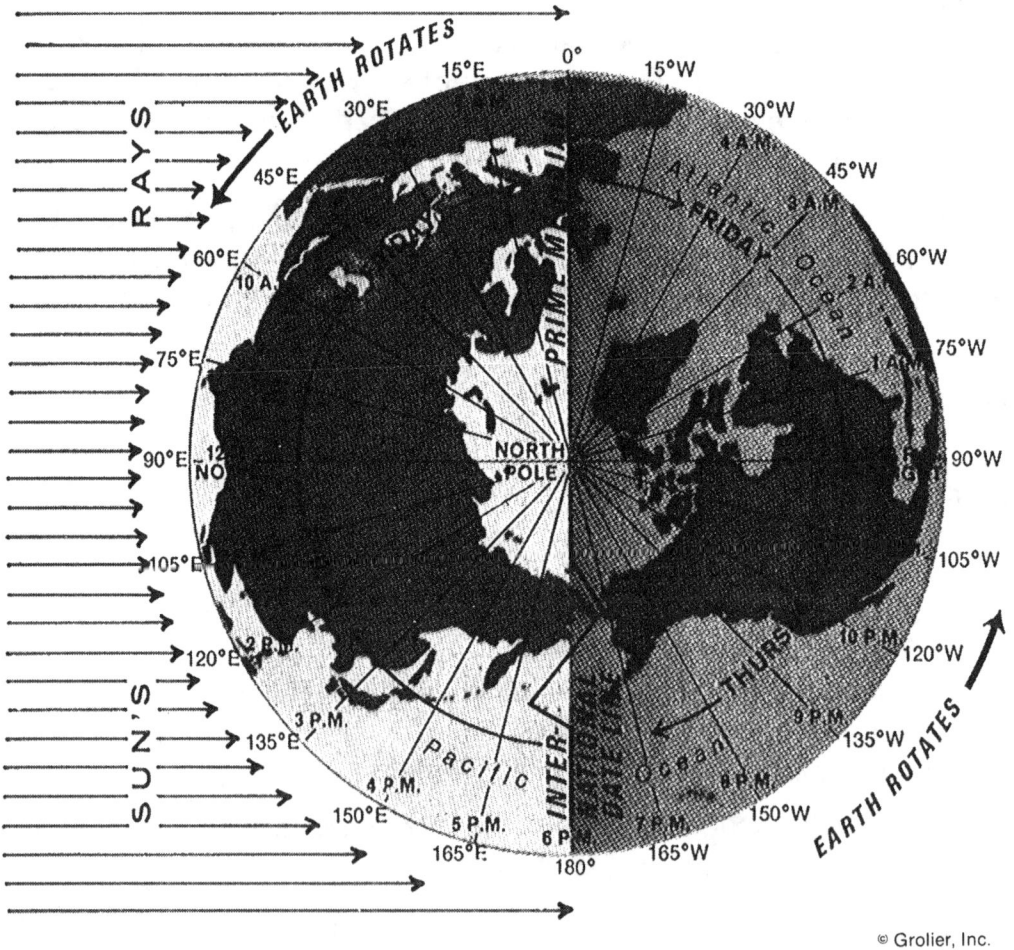

Point out that the prime meridian is in the middle of one time zone. Ask students to speculate on why the edges of the time zones do not exactly follow lines of longitude; the differences between the 180° meridian and the

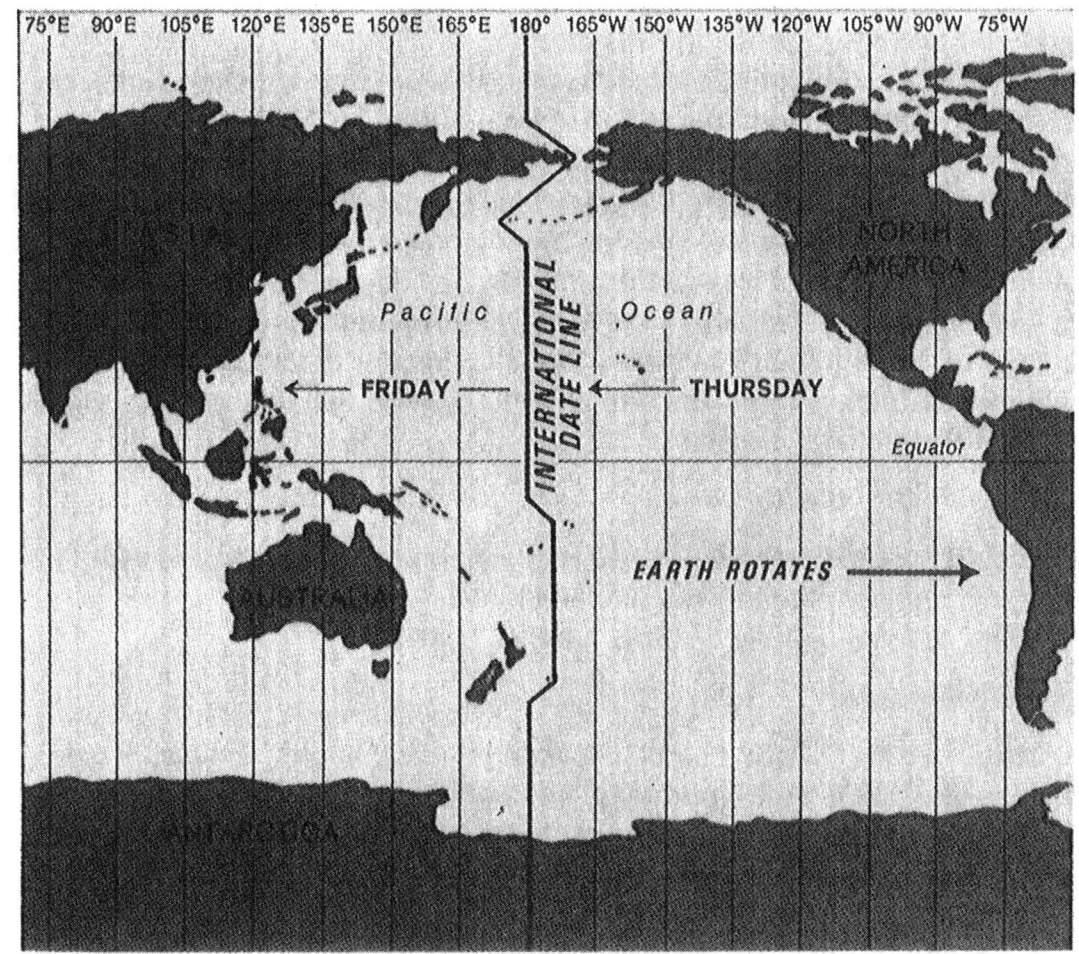

International Date Line can be used to illustrate how the edges of time zones zig and zag to include islands or parts of countries in one or another zone.

Explain that time throughout a zone will always be the same. But as we travel across zones, time changes. As we go west, time is one hour later for each time zone we cross. As we travel east, time is one hour earlier for each time zone that we cross. Give students a few simple practice problems to reinforce this notion. For example: If it is 10 a.m. Monday in Washington, D.C., what time and day is it in Sydney, Australia? (It is 1 a.m. Tuesday in Sydney.)

Follow-up: Keep the time-zone map always displayed in the classroom so that students can refer to it during current events discussions.

Additional Activities

- *Important numbers on a global grid*

Write the following numbers on the chalkboard: 360, 69, 365¼, 24, 15, 0, 23½, 66½, 25,000, 180, and 90. Ask students to explain the significance of each of these numbers to the global grid system.

Answers: 360—number of degrees in a circle; 69—approximate number of miles in one degree of latitude; 365¼—days required for earth to make one complete revolution around the sun; 24—hours required for earth to rotate once on its axis; 15—degrees of longitude per time zone; 0—latitude designation at the equator and longitude designation at the prime meridian; 23½—degrees of latitude at the tropics, as well as the inclination of degrees of the earth on its axis; 66½—degrees of latitude at the Arctic and Antarctic circles; 25,000—approximate circumference of the earth at the equator in miles; 180—highest longitude designation, near which the International Date Line runs; 90—highest latitude designation, at the poles.

- *Using latitude and longitude*

Distribute Supplements 6-4-2 and 7-4-2, which provide practice in using the global grid, and ask students to complete them.

Answer key for Supplement 7-4-2: a cruise ship.

- *Mystery message*

Write the two lists of grid points below on the chalkboard. Ask students to locate the grid points on a world map, numbering each location as they find it. Students should then connect the numbers in each section to spell out a mystery message. (Answer: the U.N.)

Part 1	**Part 2**
1. 60°N, 150°W	1. 30°S, 30°E
2. 15°N, 150°W	2. 30°N, 30°E
3. 30°S, 150°W	3. 60°N, 30°E
4. 30°S, 105°W	4. 30°N, 60°E
5. 30°S, 60°W	5. 0°, 90°E
6. 15°N, 60°W	6. 30°S, 120°E
7. 60°N, 60°W	7. 60°N, 120°E

- *Estimating distances using latitude*

Because latitude lines are always the same distance apart (approximately 69 miles), they can be used to estimate north/south distances on a map. To reinforce this idea for students, ask them to estimate distances in miles between two locations. This can be done by multiplying 69 by the number of degrees of latitude between the two locations. Remember, only north/south distances can be calculated in this way; east-west distances are measured by longitude lines, which are not equidistant at all points.

The following are some questions that could be answered by using latitude to estimate approximate distances in miles:

- How many miles is Winslow, Arizona, from Bozeman, Montana?
  (Winslow: 35° N; Bozeman: 45° N = 690 miles)

- How many miles is New Orleans, Louisiana, from St. Joseph, Missouri?
  (New Orleans: 30° N; St. Joseph: 40° N = 690 miles)
- How many miles is Charleston, South Carolina, from Pittsburgh, Pennsylvania?
  (Charleston: 70° N; Pittsburgh: 40° N = 2,070 miles)
- How many miles is Tennessee's southern border from its northern border?
  (southern border: 35° N; northern border: 37° N = 138 miles)
- How many miles is Wyoming's southern border from its northern border?
  (southern border: 40° N; northern border: 44° N = 276 miles)
- How many miles is the southernmost tip of Texas from the Canadian border directly north of Texas?
  (Texas: 26° N; Canadian border: 49° N = 1,587 miles)
- How many miles north is Minneapolis from Los Angeles?
  (Minneapolis: 45° N; Los Angeles: 34° N = 759 miles)
- How many miles south is Miami from Seattle?
  (Miami: 26° N; Seattle: 48° N = 1,518 miles)

Interpreting Maps

Several simple strategies can help students begin to interpret maps. One is to expose them to a variety of maps. Another is to provide a specific purpose for using each map. A good way to encourage student participation in map interpretation is to ask a series of questions such as: What is the title of the map and the orientation? What symbols are being used? What are the specific data on the map? What conclusions can be drawn from the data?

Activity 1-5-2 *Different kinds of maps*

Purpose: To distinguish between general-reference and special-purpose maps and to recognize that a map's title tells its purpose

Materials:

1. copies of Supplement 1-5-2
2. copies of a U.S. map from the Map Masters section in the appendix
3. reference materials (almanac, encyclopedia)

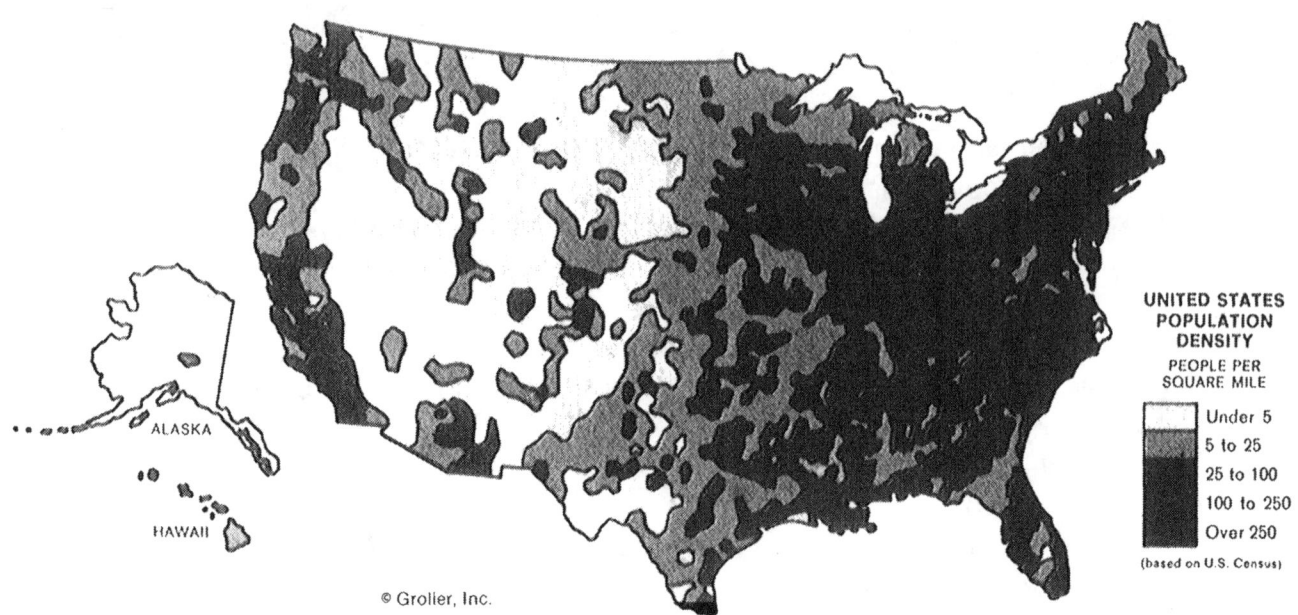

UNITED STATES
POPULATION
DENSITY

PEOPLE PER
SQUARE MILE

Under 5
5 to 25
25 to 100
100 to 250
Over 250

(based on U.S. Census)

© Grolier, Inc.

Procedure: Ask students to make a list of the many different kinds of maps. Explain that there are two general categories of maps: **general-reference maps**, which show several kinds of information (boundaries, rivers, lakes, cities, roads), and **special-purpose or thematic maps**, which show one special kind of information (weather in a country, where certain types of plants grow, how many people live in various places and so on). The map title usually will tell you the purpose of the map.

After you have shown students several examples of thematic maps and general-reference maps, divide the class into several small groups. Have each group choose a topic for which they will draw a thematic map of the United States. They can use the outline map from the appendix as their base map. Remind students to choose titles that show the purpose of the map and use reference materials for help in creating their maps.

Follow-up: Make a bulletin board display of as many different kinds of maps as possible. Encourage students to bring in maps from newspapers, magazines, and other media.

For a follow-up activity on special-purpose maps, you might suggest that students research a particular topic and then highlight the data on a map. For example, you might ask them to

(1) Make a list of the tallest buildings in the United States.
(2) Create a symbol to indicate that the building is taller than 800 feet.
(3) Show the location of the buildings on a U.S. map.

Distribute Supplement 1-5-2 and ask students to complete it.

Activity 2-5-2   *Map: fact or fiction*

Purpose: To understand that a general-reference map is the best type of map to use in answering questions that refer to location and direction

Materials:

1. copies of Supplement 2-5-2
2. detailed world map or atlases

Procedure: Remind students of the difference between general-reference and special-purpose maps. Distribute copies of Supplement 2-5-2 and ask students to name the kind of map that they would use to find the information listed. (general-reference map) Point out that because general-reference maps show several kinds of information, they are often most helpful in answering questions that refer to location and direction.

Ask students to complete the supplement. (All the statements are true.)

Follow-up: Provide groups of students with atlases and have each group make up a list of "I Spy" clues. For example, one team might come up with the following clues:

- "I spy the only place in the United States where four states meet."
- "I spy the country that produces the most beef."
- "I spy two countries in the Southern Hemisphere where English is the official language."

Teams should exchange clues and identify the kind of map that is needed to answer each clue; they should then use the atlases to answer the questions.

Activity 3-5-2 *Reading road maps*

Purpose: To understand that road maps are helpful in planning trips and locating places

Materials:

1. set of state road maps (one for every two students)

Before beginning this activity, select five pairs of locations in your state and write them on the chalkboard. For example: Chicago and Champaign-Urbana, Springfield and Cairo, Starved Rock State Park and Freeport

Procedure: Distribute the state road maps and have the students use the direction arrow or compass rose to orient the map. Then have students point out the various symbols used for roads, cities, towns, points of interest, and public recreation areas; the map scale; the alphanumeric grid used; and any inset maps. For preliminary practice, ask students to locate particular places in the state and identify what direction they are from a central point.

Next, ask student pairs to plan the best route between each pair of locations you have listed on the chalkboard. Depending on the students' skills, you can make the task more challenging by requiring them to calculate distances, the hours of driving to reach the destination, the tolls required, and the anticipated gasoline costs (based on a fixed index of miles per gallon and a set cost for a gallon of gasoline).

Allow students to compare their routes with those planned by other student pairs.

Follow-up: A writing project is a very good way to reinforce the skills gained in the above activity. Ask students to write paragraphs describing what they would see as they traveled along the routes they planned. They could make a chart, listing such things as cities/towns, airports and other means of transportation, historic sites, monuments, and so on. Or they might make a wall chart showing the highway symbols they encountered on their route.

Activity 4-5-2   *Setting up a weather station*

Purpose: To know how to read and interpret weather maps

Materials:

1. U.S. maps
2. weather maps
3. thermometer, weather vane, rain gauge, barometer
4. chart of weather map symbols (see chart below)

Before beginning this activity, duplicate a U.S. map from the Map Masters section of the appendix and several weather maps obtained from newspapers or the National Weather Service (National Meteorological Center, World Weather Building, Room 307, Suitland, MD 20233).

A Daily Weather Map

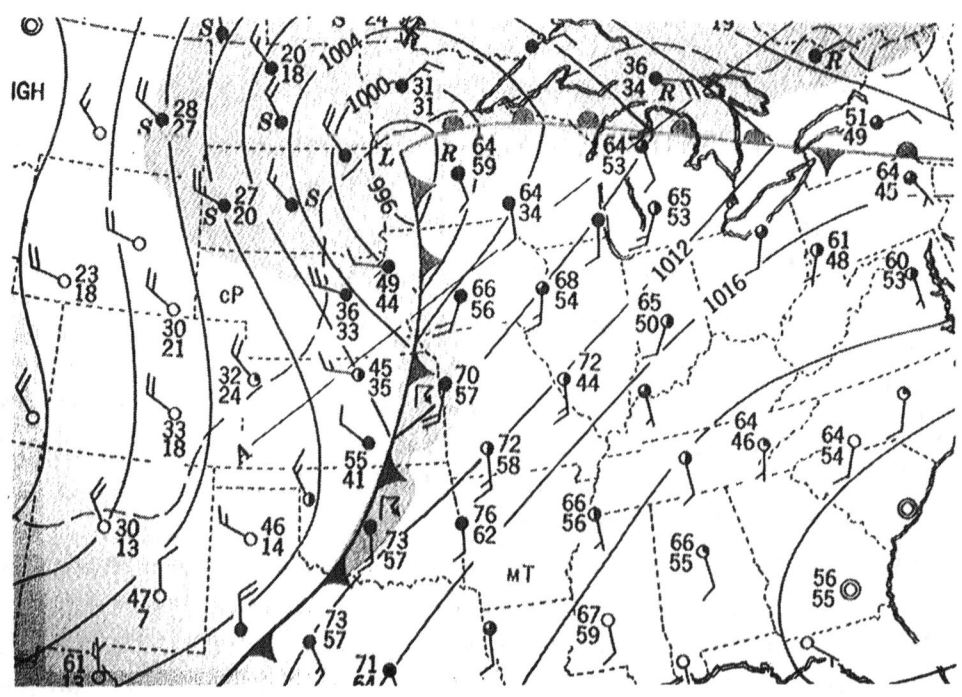

Procedure: Distribute Supplements 4-5-2 a, b, c, d on weather symbols and weather maps. Discuss the information they provide. Ask students to bring in weather maps from the newspaper for several days; some students may also be assigned to watch or listen to radio or television weather reports and make notes on the kind of information presented. Each day, discuss the results.

When students have gained some experience with reading weather maps, have them set up a classroom weather station to gather daily weather information. Ask individuals or groups to record data at set times each day. The data can then be recorded on log sheets. Data can be obtained by the following methods:

- recording the temperature from a thermometer on the sill or frame of a classroom window
- observing cloud types and cloud cover
- observing precipitation, wind direction, and speed
- calling a weather station daily

When students have produced a series of weather charts, conduct a class discussion of the patterns they show. Analyze the causes and effects of various weather patterns. Have students continue collecting and mapping their data, but also assign them to begin making weather forecasts for the local community. Periodically discuss whose predictions are most accurate and why.

Follow-up: Have students gather daily weather data and compare it with previously recorded weather. They could choose data from the same date 10 years, 50 years, and 100 years ago. Ask them to make graphs or charts comparing the data. Actual weather reports could be presented by students in language arts class, or a "Did You Know" weather booklet could be compiled by the class.

Activity 5-5-2 *Making inferences*

Purpose: To identify where a story occurred by using geographic clues and maps

Materials:

1. atlases
2. Geographic Regions maps from an atlas or encyclopedia
3. copies of Supplement 5-5-2

Procedure: Give each pair of students a Geographic Regions Map of the United States. Ask them to study it and make a list of the different kinds of information it provides. Conduct a class discussion of the results.

Distribute Supplement 5-5-2 and have them complete it, or have them listen carefully as you read the following stories. Have them find words in the story that will help them identify the geographic locations.

Anaki could hear the rain hitting the **bark walls** of her family's wigwam. Slipping on her **deerskin moccasins**, she walked across the **pine needle carpet** to look outside. She hoped it would stop raining before it was time to work in the fields that her family had cleared in the **forest**.

Today they would be **picking beans and corn. Potatoes and squash** also grew in the fields. Although Anaki liked working in the fields, she did not like getting wet. Just then, she saw her older brother walking through the village toward their **wigwam**. He carried his **spear** and several **fish**. Suddenly Anaki felt hungry.

Ask students: In what region of the United States might you find these things? (in the northeast) When might the story have occurred? (1600s or 1700s)

Repeat the exercise with this modern story, which most likely took place in Arizona or New Mexico.

Lisa's alarm clock rang at 5:00 A.M. Even though the **sun** had not risen, the **desert** was already warm. Lisa knew she would be hot when she finished delivering papers on her paper route. She was glad there were **no hills** on her route.

Pulling on her jeans and a T-shirt, Lisa ran downstairs for a quick bowl of cereal and a glass of juice. The **air-conditioned** house was quiet. Her mother would not be up for another hour. She did not have to leave for her job at the **solar energy company** until 7:00 A.M. Lisa ran down the front steps of their house and jumped on her bicycle. She was off!

Follow-up: Place a large outline map of the United States on the bulletin board. Divide the class into seven groups and assign each group a region of the country. (Refer to Supplement 2-4-3.) Have the groups study the atlases to develop a description of the region assigned to them. Ask them to cut out photos from magazines to illustrate their region and attach them to the map on the bulletin board.

Additional Activities

- *Jeopardy map board*

Obtain a large pegboard. Place twenty-five golf tees or small hooks in the holes. Ask students to develop geography questions that can be answered by referring to the classroom wall map. Have them write the questions on 5" x 8" file cards. The questions might be organized as follows:

- geographic terms
- resources and products

- key cities or capitals
- geography and history
- famous places

Evaluate the relative difficulty of each question and assign a point value (10, 20, 30, 40, or 50). Put the point value on the back of each card and attach each one to the pegboard with a golf tee or a hook. Keep the side showing the point value face-up.

Conduct the game like the "Jeopardy" television program, but have students compete in teams rather than individually. Assign a student judge or a three-student jury to keep time and rule on the accuracy of players' responses.

- *Identifying a region*

Draw a blank outline map of a small area of a region in the United States that shows only rivers, lakes, and various other natural features. (Choose areas that have, in fact, been developed.) Do not identify the region. Have students draw in cities, railroads, highways, etc., where they believe they should be located. Do not let them refer to other materials while they are making these hypotheses. Discuss the reasons for their decisions. Then refer to an actual map of the area, comparing actual locations of cultural features with students' answers.

- *Mystery location*

Every day, ask one student to select a mystery location on the world map, write the name on the chalkboard, and cover it with a large piece of paper. Throughout the day have the student write clues on the paper to help the other students guess the mystery location. When someone thinks he/she has discovered the mystery location, he/she should write the guess on an index card and place it on the chalkboard tray. At dismissal time or early the next morning, the student who selected the mystery location should reveal it and read the guesses that were made.

- *Writing a travelog*

Ask each student to select a city, state, or country and research the following information: location, size, natural features, population, food, government, economy, and tourist attractions. Then, acting as travel guides, the students should write a travelog and read it into a cassette tape recorder. The recorded reports can be played back later as the class follows each stop on the featured itinerary.

Making Maps

Mapmaking is a good way to reinforce what students have learned about map characteristics. Opportunities for mapping are endless, limited only by the restrictions we place upon our own imaginations. Whether students draw maps or build them from various materials, they are creating the essential tools for using direction and locating places.

The following checklist will be helpful in guiding students as they begin to make maps. Have them refer to it before they begin.

A Map Checklist

1. What is the purpose of your map?
2. Do you have all the information you need to draw the map? If not, where or how can you get it?
3. Have you decided on a scale for your map? On what size paper will you draw your map?
4. How will your map be oriented? What direction will be at the top?
5. Have you created symbols for all the kinds of information you want to show on your map?
6. Do you need a grid for your map?
7. When you have finished your map, give it a title and include a legend, scale, and compass rose.
8. Sign and date your map.

Constructing a Playground Map

If your school playground has a large asphalt or cement area, you might consider getting permission to create a giant map in this area. Once you have administrative approval, select an out-of-the-way area. Then form student committees to plan and complete the project. Committee tasks might include site preparation, acquisition of materials, map enlargement, chalking in (or painting) the map.

The following ways of enlarging a base map can be used.

1. Draw a grid of numbered one-inch squares on a base map. Duplicate the same numbered squares on the playground, increasing the scale of the outdoor map to the size desired (1″ squares = 1′, 2′, 3′ squares). Use tape measures to help you draw squares accurately in chalk. Refer to the line work within each square on the base map when enlarging them on the playground map.

2. Use an overhead projector or opaque projector to enlarge your base map onto large sheets of paper that have been taped together. Trace the outline of the map with a marker. Punch holes along the map lines at close intervals. After placing the enlarged map on the playground, mark each hole with chalk. Then remove the map and connect the chalk marks to form the map outline.

3. Repeat one of the above procedures for enlarging the base map onto large

sheets of paper. But turn the sheets over and lightly moisten the back of the map outline. Then coat the underside of the map with baby powder or dusting powder. Immediately invert the map and draw over the outline with a large crayon, pressing very hard. Remove the map and paint over the powder lines.

4. Using chalk, make a series of measured-to-scale reference points on the concrete. By connecting the points, you will be drawing the outline of your map free-hand. After corrections are made, the map can be painted.

When the map has been completed, it can be used for a variety of purposes. Using chalk, students can draw or write in names of cities, products, major interstate highways, rivers, mountains, lakes, or famous monuments. It can be used for practice in using directions and finding locations. Students will have fun creating a variety of map games.

Making Three-Dimensional Relief Maps

Three-dimensional relief maps can be made from moldable mixtures available either at home or in school. To form the base of a large classroom relief map, use a quarter-inch plywood board, fiberboard, or heavy-duty cardboard. Draw, trace, or project the map shape onto the base at the scale desired. Drive some protruding tacks or small nails through the underside of the base to help secure the molded material to the base when you are ready to apply it. You might also add a coat of glue to the base before applying the molded material. Keep a detailed relief map, similar to the three-dimensional map you will be making, alongside the base of your map. It will help you determine proper elevations.

To form the molding material for your map, use commercially available clay or natural clay mixed with water. Other molding materials can be made with the following mixtures, which can be colored with food coloring during the preparation process or painted after they have dried:

- no-cook salt clay:
  Mix 2 cups flour, 1 cup salt, and 1–2 tablespoons vegetable oil. Add water a bit at a time until dough is pliable. Store in an airtight container or plastic bag until ready for use.

- stove-top play dough:
  Mix 1 cup flour, ½ cup salt, and 2 tablespoons cream of tartar. Add 1 cup water and cook over medium heat for 3–5 minutes. This mixture will not look workable until it has cooled. Store in an airtight container or plastic bag.

- play clay:
  Mix together 1 cup cornstarch, 2 cups baking soda, and 1¼ cups of water. Bring ingredients to a boil, stirring constantly until a dough-like consistency is reached. Remove and

cover with a damp cloth until cool. Knead and model the mixture; then allow to dry overnight.

- salt ceramic
  Mix 1 cup of table salt, ½ cup cornstarch, and ¾ cups water in a double boiler. Cook over low heat, stirring constantly until thick and dough-like. Place mixture in wax paper or aluminum foil. When cool, knead as you would bread dough. Keep refrigerated until ready to use.

In constructing a three-dimensional relief map, students may try to accomplish too much in one class period. Be sure to allow the base material to dry (at least overnight at normal temperatures) before painting, coating with a sealer, or adding additional information.

Making Papier-Mâché Globes

Making papier-mâché globes can be fun and instructional, but messy! To undertake this project, you will need round balloons, a large supply of newspapers that have been cut into strips, and thick wallpaper paste mixed as per instructions on box. The newspaper strips should be soaked in the wallpaper paste and applied, a layer at a time, to the inflated balloons. (Students will need to add at least two layers and perhaps as many as six.) Older students may preplan how they will lay out the landforms on the globe and place clay, or balls of tissue paper, under the newspaper strips to indicate elevations. When the layers of paper have dried, release the air from the balloon. Then have students paint oceans and landforms onto the globes.

Ingredients for Making Innovative Maps

Bags When constructing a floor layout map, you might use different-sized paper bags to represent buildings. Bags can be made to stand by cutting slits in the sides and stapling or pasting the flaps to a cardboard base. Windows and doors can be drawn on the bags. Cardboard roofs also can be added. When bags are cut, stuffed, and painted, they make nice office buildings, barns, or neighborhood communities.

Balloons Inflated balloons are handy for drawing the outlines of the continents, the prime meridian, international date line, equator, tropics, Arctic circles, poles, world trouble spots, and historical routes.

Balls When you need to teach concepts relating to earth movement, balls are helpful tools. Various size balls (from Ping Pong and golf ball size to volleyball or beachball size) can be marked or painted and used to demonstrate earth's rotation, revolution, relationship to other planets, equinoxes, solstices, and day and night.

Bed sheets Have students bring old bed sheets. Ask students to use colored markers to draw special-purpose maps on them.

Boxes and other containers Inverted boxes and other containers (plastic flower pots, cups, dishes, pans, pots, cans) can be used to construct three-dimensional maps and a variety of historical and geographic models of cities, towns, villages, ranches, etc.

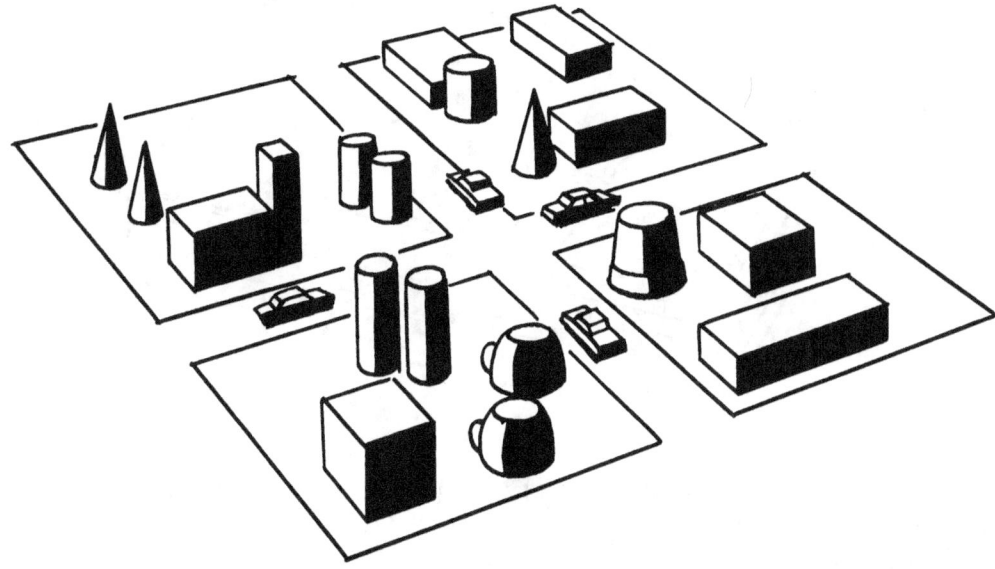

Crayon resist The crayon resist process makes a decorative map. Have students color a piece of paper with several layers of crayons. Then cover the surface with a coat of black tempera paint mixed with glue or soap to make it heavier. Use a nail or dried-out ballpoint pen to scratch lines on the coating of paint, creating the map design. The sections of the underlying crayon colors will be exposed to form the map outline.

Foil Obtain sheets of heavy aluminum foil or thin copper sheets. Have students use a ballpoint pen to trace the reversed outline of a map onto the foil or copper sheets. The tracing will create a raised effect on the sheets. Later, the embossed maps can be painted, matted, and framed. This makes a nice gift for parents or friends.

Food Various food items can be used in mapmaking also. The making of cakes, large cookies, salads, and gelatin molds in the shape of states or countries is a fascinating way to create maps. Colored frosting, placed at specific locations, can symbolize rivers, cities, mountains, etc. Alphabet letters found in cereal can be used to spell place names.

Plastic overlays and laminations Have students draw and color state, regional, or national maps and glue them to a stiff board backing. Then cover each map with a plastic overlay, taping the edges of the overlay to the backing. The map will be firmly set between the top and bottom pieces and can be used for adding more data with water-soluble pens or grease pencils. Use a large clear plastic tablecloth to cover large physical, political, or outline maps. (All the above items can also be laminated if a machine is available in your school.)

Sand Fill a low-sided, but relatively large, sturdy box with sand and place it on a table. Have students take turns shaping the sand into geographic features. Dampen the sand periodically to keep it thick and moldable.

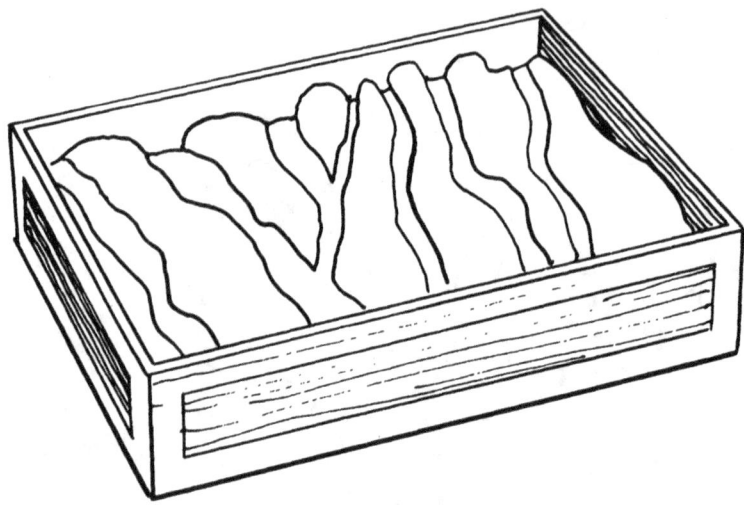

The sandscape might include students' interpretations of mountains, valleys, villages, cliffs, peninsulas, islands, etc. Label the features correctly and put in additional figures or toys to complete the scenes.

Sand casting Place wet sand in a shallow pan or box and make a mold of a map outline, one-half inch deep and about as wide as a large cookie. Pat the mold with fingers, spoon, or stick until compact and level. Ask students to add small stones or marbles for city locations. Then pour mixed plaster of paris into the mold until the plaster reaches the top of the sand mold. Place a toothpick in the center of the mold. Allow the mold to set for a half hour. Then pull on the toothpick, lifting the sand casting out of the mold. Brush away the excess sand from the plaster casting and paint your sand-cast map.

Soap bars Use large soap bars for carving map outlines. Remove the soap outside the map outline, thus raising the shape of the map on the bar. For larger maps, join two, three, or four bars together with tape. Then carve or emboss as one piece.

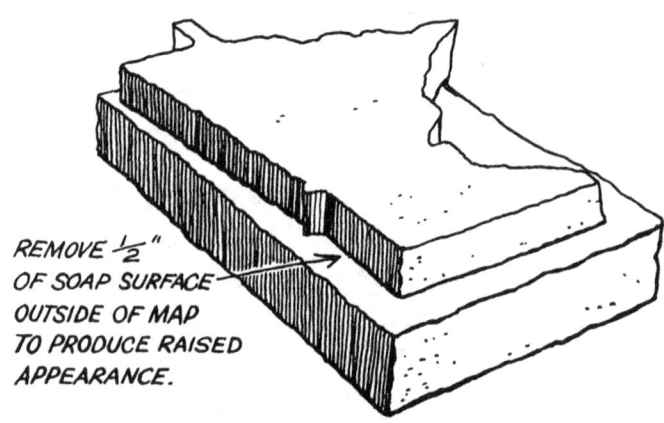

REMOVE ½"
OF SOAP SURFACE
OUTSIDE OF MAP
TO PRODUCE RAISED
APPEARANCE.

Travel brochures Readily available travel brochures, travel posters, and miscellaneous advertisements are useful for enriching map construction and studies of cities, states, and countries. They can be cut out and applied to maps or included as materials in a travel bureau area of the classroom. To obtain materials, contact tourist offices, chambers of commerce, airlines, and local travel agencies.

Wallpaper sample books Obtain discontinued wallpaper books from a decorating or home center store. Have students use the back of each sample to draw thematic maps of a specific place. Ask them to flip the pages as they make oral presentations and commentaries about their maps. Set the book on an easel for permanent display.

Wire hangers Have students make mobiles out of cardboard or colored tagboard, string, and wire hangers. The hangers can be used for hanging the outline map and each separate map cutout (strings can also be used for this purpose). Hang the cutouts according to political divisions and attach the entire mobile to the ceiling.

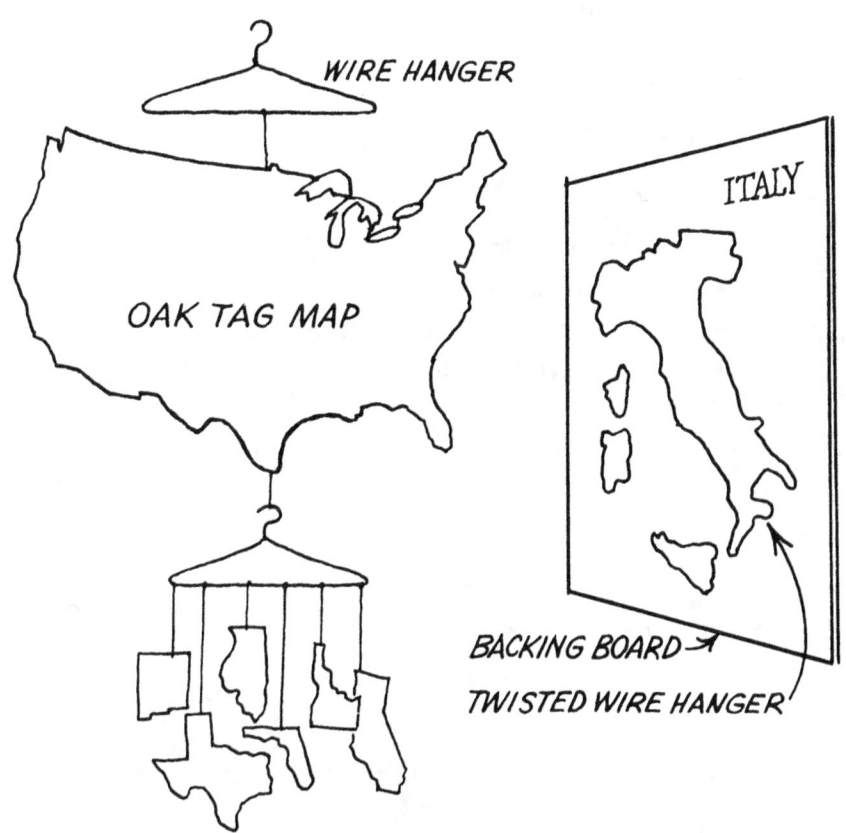

Wood Under adult supervision, students can burn, scratch, or chisel an outline of a map into the surface of a piece of soft wood. Then they can paint the crevices and sandpaper the surface thoroughly and shellac it. Students who have the opportunity to attend woodworking classes can make map breadboards out of large square or rectangular pieces of wood (one inch thickness or more). First they should trace the outline of a country or state on the board and cut it out with a jigsaw or band saw. After they sandpaper the surface, they will have a gift for family or friends.

Additional Mapmaking Materials

The following materials also can be used to make creative maps:

- building blocks, plastic interlocking pieces
- cardboard and corrugated paper cartons
- carpeting, floor tiles, linoleum
- shelving paper
- fabrics
- felt or flannel
- foam rubber
- newspapers/magazines
- paper/plastic plates
- photographs
- pipe cleaners
- dried plants, seeds
- ribbons
- screen, styrofoam
- straws, shoe laces, spaghetti, string
- tiles, metal, wood, plastic pieces
- triangles, plastic lids, covers

Part THREE

Cross-Curriculum Activities

Part THREE

Cross-Curriculum Activities

The activities in this chapter, all of which are adaptable for use at several grade levels, are organized around topics that commonly serve as course organizers in elementary and middle school social studies. These include the following:

- Using Maps at Home, at School, and in My Neighborhood
- Using Maps to Learn about Our Community
- Using Maps to Learn about Our State
- Using Maps to Learn about U.S. and World Geography
- Using Maps to Learn about U.S. History
- Using Maps to Learn about Current World Events

While this chapter focuses on social studies, the activities will also reinforce math, science, art, reading, language, and thinking skills. For example, maps will help students locate the setting of stories they may be reading in language arts classes. And knowledge of how to use grids will be helpful in math, science, and art classes.

Using Maps at Home, at School, and in My Neighborhood

The home is a child's first learning environment. From there, he or she is introduced to the neighborhood and later to the school. Thus, frequent use of maps to teach about the home, school, and neighborhood will provide early motivation for further skill development. While maps of hypothetical homes, schools, and neighborhoods can be useful in providing opportunities for practice, using maps of students' own homes, schools, and neighborhoods is especially important. It reinforces the concept that maps represent real places. The following activities involve creating and using such maps.

Activity 1-1-3 *Making simple floor plans*

Purpose: To understand that maps show where things are located, and that directional words, such as *beside*, *across from*, and *along*, can be used to describe locations. (More advanced students will be able to use cardinal directions.)

Materials:

1. open-top dollhouse or house constructed from a large cardboard box
2. toy furniture, appliances
3. tagboard cutouts or magazine photographs of actual furniture
4. large teacher-drawn floor plan of a house (see sketch below)

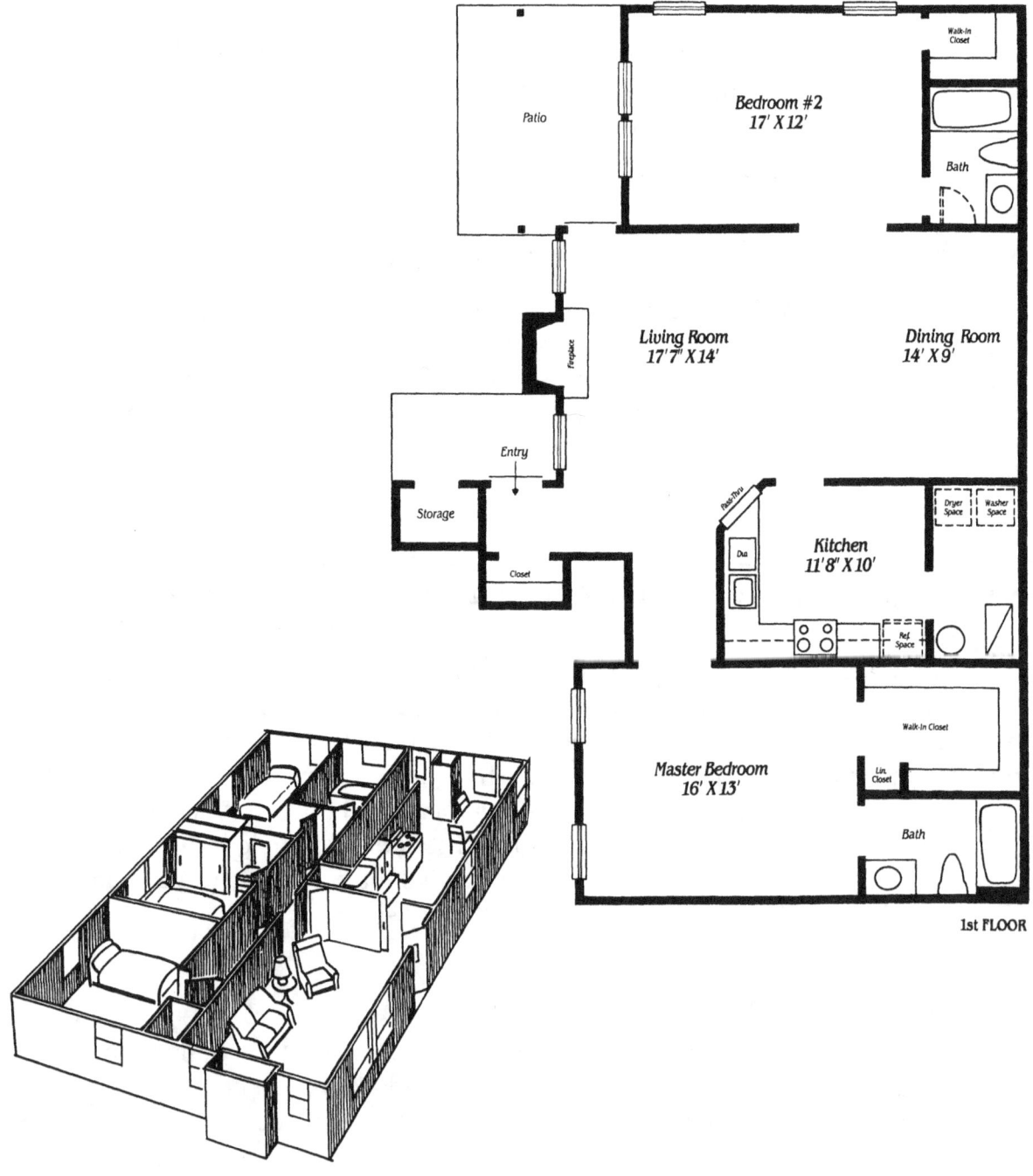

Procedure: Present the model house. Talk about the layout and the function of each room. Explain that the students will have the opportunity to decorate the rooms.

Distribute the toy furniture, giving one item to each pair of students. While one student places the item in the house, the other can describe where the item is being placed.(e.g., The couch is next to the wall.)

When all the pairs have placed their items in the house, present the floor plan. Discuss in what ways it is like a map.

Ask students to identify the rooms on the floor plan and label them appropriately. Then give each pair of students a tagboard cutout or magazine picture corresponding to an item of furniture in the model house. Have the students take turns placing the cutout in the correct location on the floor plan.

When all the students have placed their cutouts on the floor plan, ask them to choose a good title for it.

Follow-up: Encourage students to draw maps of their own homes. If they live in homes with more than one story, suggest that they concentrate on the ground floor. (See Supplement 1-1-3.) You may need to remind them that the maps do not have to show everything. If they want to make detailed maps, have them concentrate on only one room such as the family room or their bedrooms.

Activity 2-1-3 *Making seating charts*

Purpose: To understand that pictures, as well as shapes, lines, and colors are sometimes used as symbols to represent real things and to reinforce use of directional words such as *near*, *far*, *in front of*, *left*, and *right*. (More advanced students will be able to use cardinal directions.)

Materials:

1. drawing paper and coloring materials
2. large sheet of tagboard or empty bulletin-board space
   (This will be used for the layout of the classroom.)

Procedure: Prepare the tagboard or bulletin board for the student activity by drawing the classroom door in its correct location; also place a picture of yourself at the spot where your desk would be.

Ask students to draw pictures of themselves or bring in school photos. Have them write their names beneath the pictures. Explain that the pictures are symbols. In this activity, the symbols will be used to show where students sit in a classroom.

After students have taken turns putting their pictures in the correct location on the tagboard or bulletin board, ask them questions such as: "Whose desk is behind Charlie's?" "Who sits on Jaleesa's right?" "Which person in the third row is farthest from the door?"

Follow-up: Students might like to make a map of the school, using student-drawn pictures or photographs of school personnel. Before starting the project, you might tour the school with them so that they can observe various locations, sketch their layouts and, perhaps, even take photographs of key people.

Activity 3-1-3 *Using symbols to represent weather and the seasons*

Purpose: To use symbols to represent weather conditions and seasonal changes

Materials:

1. a variety of pictures (drawn by the students or cut out of magazines) that represent weather conditions and the changing seasons
2. a drawing of the school and surrounding neighborhood (Drawing may have been made by students during a previous session and selected for this activity.)
3. wire hanging over the drawing of the school/neighborhood (see illustration)

Procedure: Review the purpose of symbols. Select four symbols to represent the seasons. Attach them with clips to the wire above your school picture. If you have mounted the picture of your school on a magnetized chalkboard, glue small magnets to the backs of the symbols so they will adhere.

Ask students to match each symbol with a season. Do the same with the symbols you have selected for the daily weather changes. Then have the students select a symbol for the season and weather of the day. Make selecting the appropriate symbols one of the daily tasks to be conducted in your classroom throughout the year.

MAPLE AVENUE

Follow-up: Ask students to create symbols that stand for temperature. For example, they might create symbols for hot days, warm days, cool days, and cold days. Encourage them to use colors for these symbols. Mount a thermometer on the window frame so that students can refer to it to determine the daily temperature and add the appropriate symbol to the classroom chart.

Activity 4-1-3 *Interpreting safety symbols*

Purpose: To learn how to interpret traffic lights and other symbols that convey rules of safety

Materials:

1. shoe box and transparent red, amber, green plastic sheets
2. flashlight or filmstrip projector
3. tagboard, neighborhood map, scissors, glue, broom or mop handle
4. copies of Supplement 4-1-3

Use the above materials to construct a traffic signal (see the diagram). It will involve cutting three holes in a shoe box, covering them with pieces of colored transparent plastic, and inserting pieces of cardboard between the holes.

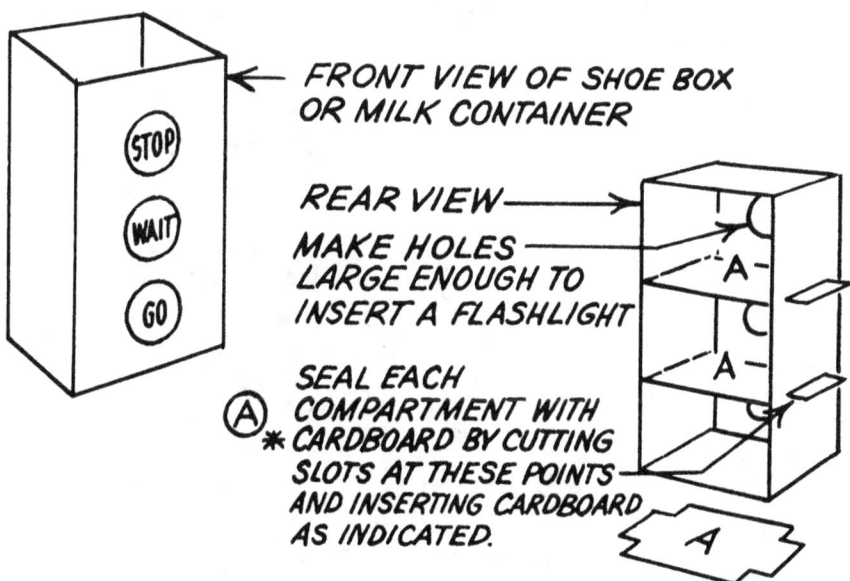

Procedure: Display the model of the traffic signal. Ask students how this safety device uses color symbols. Darken the room and use the flashlight to illuminate the signal. Use the light to simulate a crosswalk controlled by a traffic signal. Le students take turns moving across the simulated street following light changes.

Ask students to make a list of other safety symbols that can be seen in a neighborhood. Pass out Supplement 4-1-3 and discuss what each of the symbols means. Later, you might even take the students outside to locate the symbols that are found on Supplement 4-1-3.

Follow-up: For an extended activity, you can ask students to carry a small tablet around for a week and draw any new symbols that they see. Place the drawings on the bulletin board. If a large map of your community is available, post it on the board and have the students mark where they saw the various symbols. (This same activity can be done with a community map that the students have drawn.)

Ask students to look for similar uses of symbols in the school and at home. Examples would include symbols used to indicate that a product is poisonous, or symbols used to show classroom or school rules.

Activity 5-1-3 *Learning directions by observing the sun*

Purpose: To make inferences about time and direction from observing the sun and shadows

Materials:

1. a compass
2. 1 set of 3″ x 5″ cards labeled north, south, east, west (one cardinal direction per card)
3. chalk, marking pen, drawing paper
4. 10–20 direction finders (made by stapling two tongue depressors together crosswise and labeling the ends N, S, E, and W)

Procedure: Take the class outside at noontime on a sunny day. Have the students stand so that their shadows lie directly in front of them. Explain that north is in front of them, east is to their right, south is behind them, and west is to their left. Have them repeat these directions.

To reinforce the concept of direction, show students a compass and explain its function. Correlate the needle's direction with the sun and shadows. Then pass out the homemade direction finders and ask students to orient them correctly.

Give four students placards labeled north, south, east, and west and ask them to stand in the correct direction some fixed distance (such as 10 paces) from the class.

Later in the day, return to the schoolyard with the class. Using the compass, locate the sun in the west. Point out that shadows are longer now because the sun is not directly overhead. Ask students to tell you if their shadows will get longer or shorter as the sun sets. (longer)

Correlate the changing position of the shadows with a clock face. Draw a large clock face on the sidewalk, orienting noon to the north. Point out that in the afternoon the shadow points to the northeast or east.

FACING NORTH, CHILDREN POINT WITH RIGHT HAND TOWARD SUN OBSERVED IN EASTERN PART OF SKY.

9:00 A.M.

SUN OBSERVED IN SOUTHERN PART OF SKY. SHADOW IS NORTH.

11:30 A.M. OR JUST BEFORE NOON.

FACING NORTH, CHILDREN POINT WITH LEFT HAND TOWARD SUN OBSERVED IN WESTERN PART OF SKY.

3:00 P.M.

OBSERVE SUN, DIRECTIONS, AND LENGTH OF SHADOWS AT VARIOUS TIME PERIODS. OBSERVE THE SUN'S POSITION IN THE SKY AT SET TIMES DURING THE SEASONS.

The following day, return to the schoolyard as soon as the school day begins. Place a large piece of paper on the ground and have a student stand in the center of it. Trace around his/her shadow and note the time, the position of the sun in the sky, and the direction the shadow is pointing. Draw around the student's shadow several more times throughout the day, labeling each with the same information.

Another way to provide students with the opportunity to observe sun shadows is to construct a shadow stick. You can make a shadow stick by gluing a 30" wooden dowel upright in the center of a 12" square piece of wood. Following the sample sketch below, write the cardinal and intermediate directions on the wood square. Students can then take the shadow stick onto the playground at noon and orient it correctly. They can observe the movement of the shadow throughout the day, making comparisons with a clock face. A school flagpole can also serve as a shadow stick.

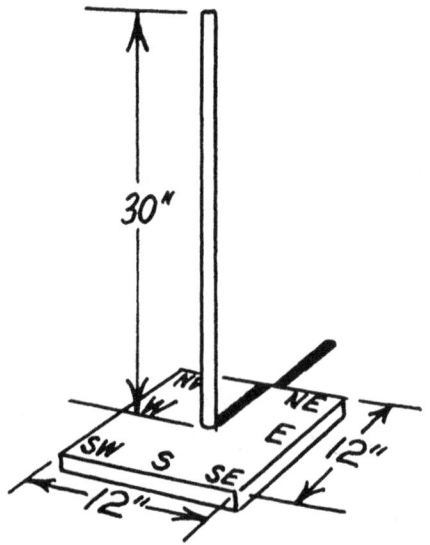

Once students have established the directions based on their outdoor learning experience, transfer their knowledge to the classroom. Post the directional placards in the appropriate locations in the classroom and refer to them when describing locations in the classroom.

Follow-up: The shadow-drawing experience could be repeated several times during the year to allow for comparisons during different seasons.

Students might draw various sun/shadow positions on large paper plates at different times of the day. They can make a miniature shadow stick by placing an upright pencil in the center of the plate and tracing its shadow on the plate. Or they could draw an outdoor scene several times, each time showing how the sun and shadows change to reflect the time of day or the season of the year.

Activity 6-1-3 *Mapping the neighborhood*

Purpose: To make a simple map of the neighborhood

Materials:

1. acetate paper
2. overhead projector

Before beginning this activity, draw a simple map of the streets within a three-block radius of the school. Provide each student with a copy. They will use it as the basis for a larger map, which they as a class will help you draw, complete with street names, houses, buildings, and neighborhood landmarks. (See sample map below.)

Procedure: Ask the students to help you make a list of all the buildings and landmarks in the neighborhood within three square blocks of the school. Put the list on the chalkboard.

Tell students that they are going to help you construct a large map showing the school and some of the other places on the list. Give each of them a copy of the blank neighborhood map you have drawn. Have them label the sides of their maps with the correct directions (or add a simple compass rose to the maps), label the streets, and draw a symbol of the school where it is located.

A map of our neighborhood

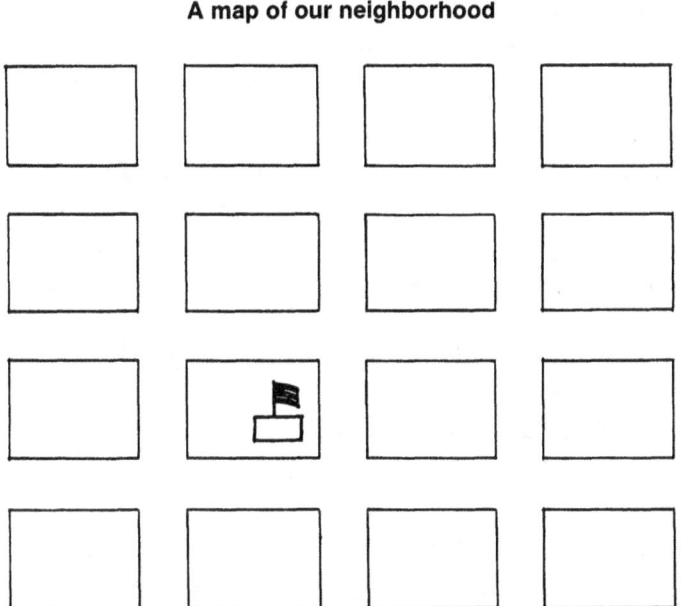

Take the students on a walking tour of the neighborhood to observe the names and locations of streets, buildings, and landmarks. Have them carry the street maps as a reference and pencil in where buildings and landmarks should later be drawn.

When you return to the classroom, draw the neighborhood map on the chalkboard or on a piece of acetate paper on an overhead projector. Have the students use their individual maps to help them tell you where all the buildings and landmarks should be drawn. When all the significant places have been drawn on the neighborhood map, have the students trace routes to and from specific areas and give directions on how to get there.

Follow-up: Encourage students who do not live in the area shown in the class map to draw maps of their own neighborhoods. If a community map is available, have students identify where the map they have created would fit into the community map.

Activity 7-1-3 *Learning about the land around us*

Purpose: To distinguish between natural and human-made features in our community

Materials:

1. magazine pictures or drawings of natural features (hills, ponds, creeks, waterfalls, rivers) and human-made features (buildings, bridges, canals, dams)
2. flash cards with the names of the items in the pictures

Procedure: Tell students they are going to be learning about special features in their neighborhood. Display the flash cards on the chalkboard and ask the children to repeat the words after you. Then hold up the pictures, one at a time, and ask the students to match each picture with the correct flash card. As they identify the items, tape the pictures and flash cards on the chalkboard in two columns: natural features and human-made features. But do not tell the students why you have placed them in two columns.

When about half the items have been identified, ask the students to tell you the difference between the pictures in the two columns. Develop a definition of natural and human-made. Then ask the students to categorize the remaining items as natural or human-made and match them with the flash cards.

Follow-up: Have students create symbols for the various features shown in the pictures. Encourage them to add more natural and human-made features to the collections you have created. Each student could make a map of his/her yard, using only two symbols: one for natural features and one for human-made features.

The pictures and flash cards could be placed in a learning center for practice in matching or to provide springboards for story writing.

Activity 8-1-3 *Tracing a delivery route*

Purpose: To be able to trace a route on a map

Materials:

1. a simple street map of your community obtained from the chamber of commerce, local tourist bureau, city hall, or real estate company
2. small toy truck
3. cards with the names of several community locations written on them

Procedure: Display the community map and ask students to identify familiar places. (Perhaps some students will show where they live.)

Tell students that they are going to pretend to be delivery workers. Give each student a card indicating a location on the community map. Allow them to take turns driving the toy truck to their locations, announcing the streets and directions as they travel.

Follow-up: Using the same community map, have students write directions for getting from one place to another. Then have them exchange these directions with other students and take turns announcing the directions they are following to get to their destinations.

Additional Activities

- *Designing a zoo*

Provide students with large pieces of newsprint and a variety of materials such as colored tape, boxes of various sizes and shapes, toy animal figures, straws, small rocks, twigs, and so on. Ask students to create a model of their neighborhood zoo or an imaginary zoo. They can simulate environmental settings using mirrors (lakes and ponds), rocks, stones, and various forms of vegetation. A color code could be added to each animal shelter to indicate the continent from which the animals came. When students have completed the display, they might draw maps based on the model or use the model to practice answering directional questions.

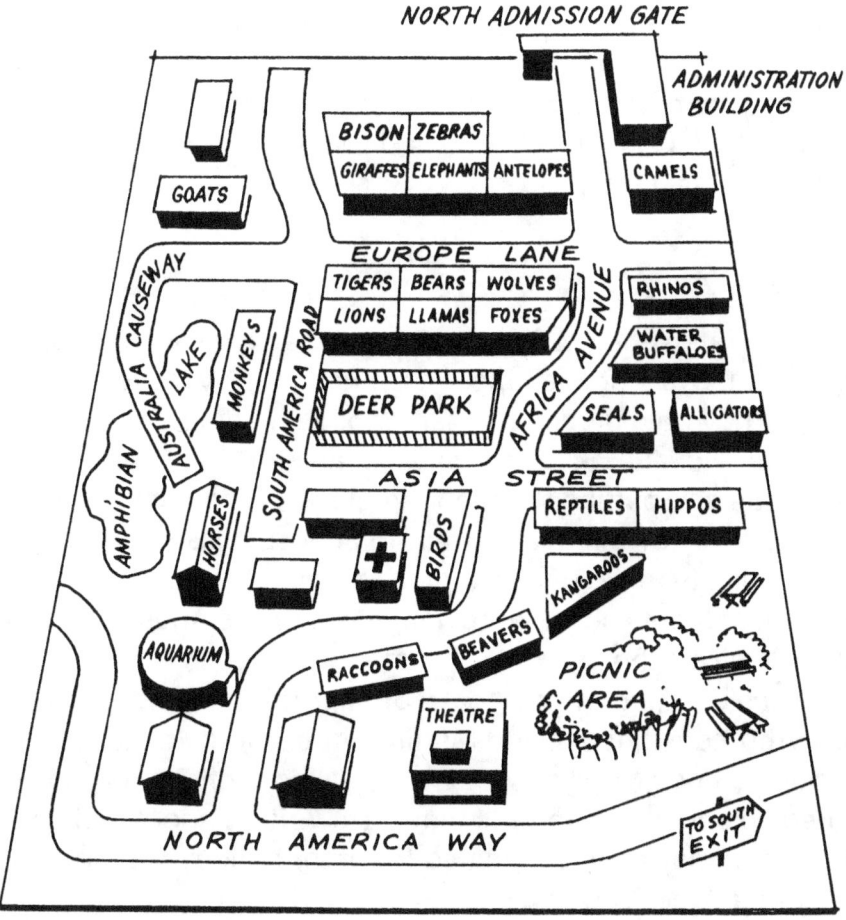

- *Simulating a supermarket in your classroom*

Have students rearrange the classroom to resemble a supermarket. Tell them to identify aisles with number or letter codes similar to those used in local supermarkets. While some students create symbols for the various categories of items (frozen foods, dairy products, meat, bakery goods, and so on), others can draw a map of the supermarket for display near the door.

Ask students to create shopping lists that reflect healthy meal planning. Then let them go through the pretend supermarket and make their purchases. You may want to provide play money so that students can practice making change.

This activity is excellent for reinforcing health and math concepts, as well as map skills.

- *Mapping interstate and foreign mail routes*

Provide each student with a map of the United States showing the state boundaries. (Use a map from the Map Masters section of the appendix.) Ask students to take the map home and, with their parents' assistance, mark the points of origin of all the mail the family receives for a week. Have the students bring in out of state or foreign postmarks cut from envelopes the family received.

In the classroom, display a large map of the United States on the bulletin board. On it have students put pins at the locations where their letters originated. You might also display the postmarks around the map; their points of origin can then be connected with yarn or string. This activity could be followed by a field trip to the local post office to find out how mail gets from place to place within the postal system.

- *Maps and holidays*

You may find it interesting to incorporate map activities into your classroom celebration of holidays. For example, on Columbus Day, have students trace Columbus's journey on a globe or a map prior to reading or listening to a story about Columbus's voyage. Students will like making small cutouts of Columbus's boats and putting them on the globe or map to indicate the direction he sailed. They might also want to affix small Spanish flags to the globe/map to indicate the areas Columbus claimed for Spain.

At Halloween, students could write haunted house stories and draw maps of floor plans to illustrate their stories.

For holidays celebrating the birthdays of important people, such as George Washington, Abraham Lincoln, and Martin Luther King, Jr., have students create a "Birthplaces/Birthday Map" on which they will identify the birthplaces of these important people and attach reports they have written about their lives.

Various kinds of maps can be created for other holidays as well, such as Veterans Day and Arbor Day.

Using Maps to Learn about Our Community

As students begin to study their community in social studies classes, they will learn concepts from geography, history, political science, anthropology, economics, and sociology. A thorough knowledge of where things are located in a community and why they are located where they are will be helpful in learning how the community functions. Frequent use of community maps is a key to developing that understanding.

Activity 1-2-3 *Exploring our environment*

Purpose: To describe geographic features and landmarks in a community and locate them on a map

Materials:

1. copies of a hand-drawn map of an interesting area of the community
2. portable cassette recorder
3. camera (helpful but not essential)

Before beginning this activity, choose an interesting area of the community in which to conduct a class tour. Explore the area in advance, noting special landmarks, vantage points, and geographic features. Draw a simple map of the area, noting each of the special features by number. Make enough copies of the map for all the students.

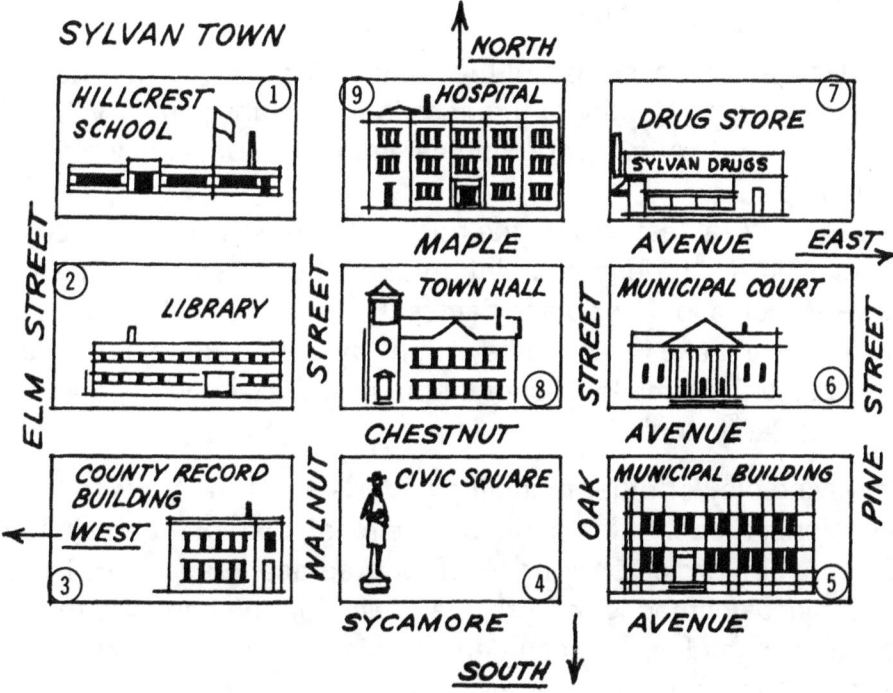

Procedure: Explain to the students that they will be taking a tour of an area of their community in order to gather information about it. Pass out the maps you have made and explain what the numbers stand for.

Begin the walk. At each numbered area on the map, stop and have students take turns recording their observations into the tape recorder. Ask several responsible students to take turns photographing each area.

After returning to school, play back the students' observations and, as a class, trace the route followed on the map. Draw a legend on the map to indicate which features were found at each numbered location. Discuss the tour. Ask students to tell you how the area resembles a similar area in another community and how the area resembles or differs from the rest of your community.

Follow-up: If photographs were taken at each of the sites of the explored area, enlarge the class map to a size suitable for posting the photographs. Have the students compare their own neighborhoods with the area explored. How are the two areas similar? How are they different?

Activity 2-2-3   *Determining place names for the natural features of the environment*

Purpose: To understand how elevation and natural features are shown on a map and how they influence place names

Materials:

1. relief map of the local community. (Older students can use a contour map.)
2. political map of the community

Procedure: Display the relief map of the community and discuss the map symbols. Help students identify differences in the height of hills, valleys, mountains, mesas, and other topographical features. Bodies of water should also be identified.

Compare the natural features on the relief map with the same places on a political map. Ask the students to point out the differences/similarities.

Next, have students look for place names that have been derived from natural features (Hill Street, Lake Street, Valley Highway, Ridge Road, Foothills Parkway, and so on). Compile a list of the names. Discuss why people name places after natural features.

Follow-up: Using clay or other modeling materials, ask students to build a three-dimensional model of the community's relief. Older students might use an overhead or opaque projector to trace a topographical map onto a large sheet of paper. Later, places named for natural features could be added to the map. When the map is completed, ask students to discuss how the natural features influenced the shape of the community.

Activity 3-2-3 *Discovering the history of our city*

Purpose: To compare a historical map of a community with a contemporary map to see how the community has changed

Materials:

1. crayons or marking pens
2. current and historical maps of your community

Procedure: Tell students that they will be learning about the history of their community by comparing a historical map of the community with a contemporary one.

Distribute the historical and contemporary maps to the students. If maps do not reflect landmarks or specific geographic features, add them to the maps. Have students do research in the library, museum, or historical society to find old photographs, newspaper articles, postcards, etc., that would document the history of the time. Place copies of these items on your class bulletin board, next to the two maps.

Then compare the two maps and discuss:
- How big was our community? What were its boundaries? What is its present size? What are its present boundaries?
- How many streets did our community have? How many does it currently have?
- What buildings/landmarks were present when the older map was made? What buildings/landmarks are present today?
- How has the community changed? What caused the change?
- How has the community stayed the same? Why?

Follow-up: Ask students to investigate the origins of place names shown on the historical map. What do the stories behind the place names tell us about the community's history?

Activity 4-2-3 *Land use in our community*

Purpose: To categorize land use areas and locate them on a community map

Materials:

1. large community map with a plastic overlay
2. several color markers

Before beginning this activity, draw boundaries around each different land use area on the overlay such as a shopping center, industrial area, farming area, residential area, public area (including government buildings or schools), and a park, national forest, or forest preserve.

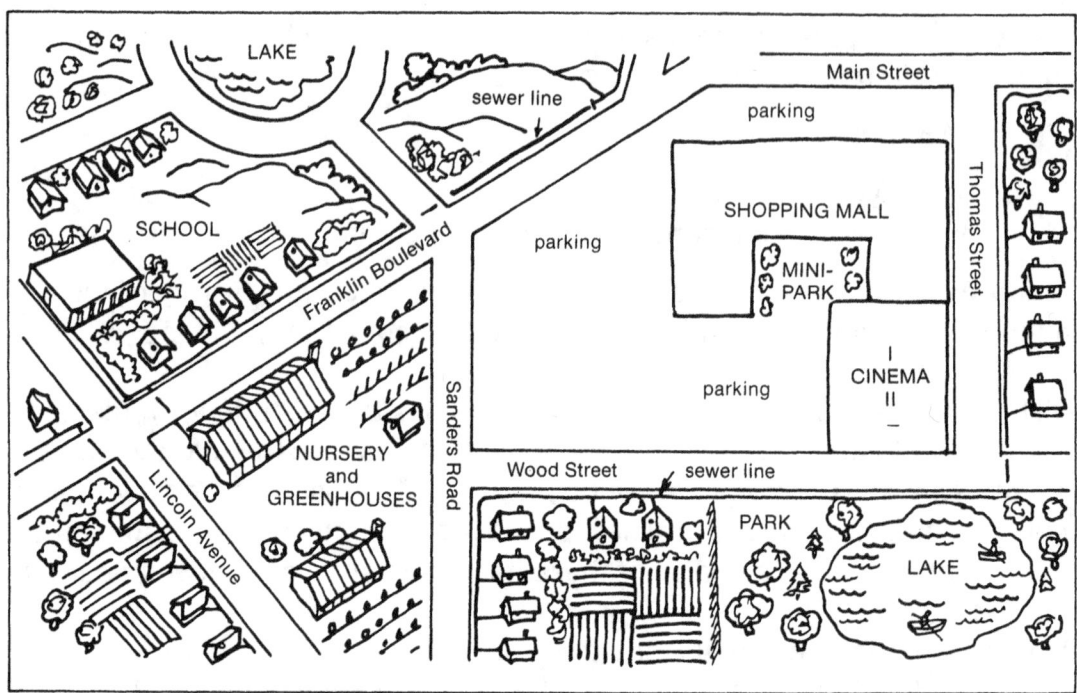

Procedure: Discuss ways in which people use land (farming, housing complex, park, and so on). Display the community map you have prepared. Point to the various outlined areas. Ask students to choose a color to represent each land use area. Make a legend for the map.

Follow-up: Provide students with a collection of old newspapers and magazines. Ask them to cut out pictures to represent the various ways land is used in your community. The pictures can be used as a visual key to the map overlay.

Activity 5-2-3 *Comparing communities*

Purpose: To compare and contrast communities

Materials:

1. copies of Supplement 5-2-3

Procedure: Distribute Supplement 5-2-3. Divide the class into small groups, giving each one several questions from the worksheet to research in the library. At the end of the week, share the answers. Have the students present their research in a report, story, or travelog. Ask them to use maps, whenever appropriate, to present their research.

Follow-up: After students have presented their investigations of the community, have them select other communities to research. Using the same worksheet, have them repeat the activity. Then compare the various communities, discussing which communities have been influenced the most by geography, which communities have the most to offer people, and which communities are most alike.

Additional Activities

- *Collecting community maps*

Challenge students to find as many different kinds of maps of your community as possible. These might include tourist maps, a map of bus routes, real estate maps, standard road maps, maps showing parade or race routes, and so on. Display the maps on the bulletin board.

- *Mapping a marathon*

After explaining to students that a marathon is a 26-mile race, run over an open course rather than on a track, have students draw their own marathon route on a community map, using symbols or arrows. Make sure that the route passes through various land use areas of the community. Have students use cardinal directions to describe the route (e.g., north to the Overland Bridge, southeast to the Twin Towers, and west to Lake Clearwater).

- *Making an economic activities map*

After students have studied the community's major products, ask them to draw symbols for these products. Have them place the symbols on a community map to show how the products are transported to other communities. If products are trucked out of the city, for example, the symbols could be placed along major highways leading out of town; symbols for products that are flown out of the city could be placed at the airport.

If your community is a suburb with no manufacturers, you might focus on services provided and the goods used in such services. You might trace where these goods come from.

Using Maps to Learn about Our State

Maps are essential in the study of state history. Because of the diverse geographic, historic, and economic nature of the states, you will need to gather information from a wide variety of state maps. These can be obtained from state government agencies, state departments of tourism, and some commercial publishers.

Activity 1-3-3 *Studying place names*

Purpose: To become familiar with place names in a particular state, relating them to historical events or to the natural features of the area

Materials:

1. political and physical maps of a state
2. reference material on state history and geography

Procedure: Tell students that learning the origin of state names will provide them with information about the history and geography of the state.

Divide the class into several small groups, assigning each group one section of the state. Ask them to find places in their particular section that were named after the following sources:

> (1) a specific occupation or economic activity
> (2) a person or group of people
> (3) a specific place in another country
> (4) a plant, mineral, or animal
> (5) a body of water
> (6) a landform

In addition, ask each group to identify at least two places that have unusual names which do not fit any of the above categories. Have the students organize their information in a table with the following column headings:

place name   origin   location   related facts

When all groups have completed their research, ask them to share the results and display the research. Then discuss what the place names reveal about the state's history and geography.

Follow-up: To supplement the above activity, students can conduct special research into names derived from Native American languages. Using a map of the United States, have students color-code sections of the map to indicate which states have the greatest concentration of Indian place names. Make a list of the place names according to tribes or language patterns and give the English equivalent of the names, if possible.

Distribute Supplement 1-3-3 to the students who would like additional work in locating place names. Have them use an atlas to find places in the United States that are named for the natural features listed there. The state names are given as clues.

Activity 2-3-3   *Constructing a mileage chart*

Purpose: To calculate distances between cities and convert the information into a mileage chart

Materials:

1. several road maps of your state
2. rulers
3. several 3″ x 5″ cards listing pairs of cities within the states

(none of the pairs should be repeated, but several large cities should be represented on each card, as should your local community)
4. newsprint or butcher paper
5. sample mileage chart

Procedure: Divide the class into several groups, giving each group a state road map and a card listing pairs of cities. Tell each group to use a ruler to calculate the distance between each pair of cities listed on the card. Each calculation should be checked by at least one other group member. Then the distance should be noted on the card.

When all the groups have completed their calculations, they should work together to construct a mileage chart. You may wish to provide the students with a sample chart, which you can photocopy from any of the state maps found in a road atlas. Or you can get one from the state office of tourism. (See sample map.)

Mileage Between Principal Cities in California

	Alturas	Bishop	Crescent City	Eureka	Fairfield	Los Angeles	Medford OR	Modesta	Oakland	Oroville	Placerville	Portland OR	Redding	Reno NV	Sacramento	San Diego	San Francisco
Eureka	309	571	84		286	680	188	360	282	252	341	412	158	377	298	807	282
Sacramento	311	273	382	298	44	382	312	67	82	66	43	586	163	133		509	90
San Francisco	367	328	365	282	47	387	367	85	8	152	138	637	218	226	90	514	
San Jose	402	364	409	325	79	347	401	82	42	185	169	675	251	259	126	468	44
Stockton	362	284	416	332	56	335	363	29	74	117	94	637	214	184	51	459	82

Follow-up: When students have completed their mileage charts, have them compare their calculations with those printed in a road atlas. You might also have them compare their calculations (based on using a ruler and a mileage scale) with the totals they get from adding the mileage numbers indicated along highways and other roads on a road map. Discuss what accounts for the differences.

Activity 3-3-3 *Making thematic state maps*

Purpose: To learn how to gather information on a specific theme and convert it into a map

Materials:

1. an 8½" x 11" outline map of a state
2. several 8½" x 11" transparent, plastic sheets
3. permanent markers
4. a state atlas and other reference materials

Procedure: Review the definition of a thematic or special-purpose map (one drawn to show a specific kind of information). Give examples. Then make a list of the kinds of information that can be shown on a thematic map of your state.

Divide the class into groups of four or five students and assign each group one category from the list. If you need other categories, use the following:

population density	*vegetation zones*	*climatic regions*
average temperature	*average precipitation*	*natural resources*
economic activities	*ethnic composition*	

Give each group an outline map of the state and several transparent plastic sheets. Show the groups how to make an overlay for the outline map that will show specific kinds of information based on their categories. They can color in certain areas of the overlay or draw pictures and symbols on it to create the thematic map.

When the overlay has been completed, group members should make at least one generalization based on the information they have mapped. For example, a group mapping average precipitation might conclude that the areas of the state west of the mountains receive more rainfall than areas east of the mountains.

Once the groups have finished their work, discuss the maps and their generalizations.

Activity 4-3-3 *Creating a state history map*

Purpose: To create a state history map with a timeline as the legend

Materials:

1. large 18" x 24" outline map of the state
2. markers
3. roll of shelf paper or newsprint
4. reference materials on state history

Procedure: Present a number of significant events of your state history, listing them on the chalkboard. Following a discussion of the events, divide the class into several groups, assigning each group an equal span of years of state history. Ask each group to select two or three important events in that period and create a symbol for each event. Have students draw their symbols on the large outline map of the state to show where the historical events occurred. For each event, the group will also fill out a card with the following information and fasten it to the bottom of the map in the order in which the event occurred:

- explanation of the symbol
- date when the event occurred
- importance of the event

When all the cards are in place, they can be used both as a timeline and as a legend.

Follow-up: Construct a "Hall of Fame" map, similar to the state map suggested on page 70. Symbols would be drawn to represent famous people from the state and placed on the map where the people were born. On an accompanying poster, have the students list and explain the symbols and attach pictures of each person, along with his/her birthdate and their contribution to state history.

Activity 5-3-3 *Making a state fair map*

Purpose: To gather information about the special features of a state and transfer that information to a map

Materials:

1. an 18" x 24" outline map of the state
2. magazine and newspaper articles about the state
3. writing materials, envelopes, stamps
4. large table

Procedure: Tell students that they are going to construct a table display, "A Mini-State Fair," showing the products and special features of their state. Each student will be responsible for gathering information about two cities or towns in the state and displaying it at the state fair table. The information will then be placed on a state map.

Before beginning this activity, you may want to make a list of all the cities and towns in the state that have contributed to its economic growth, either by way of supplying natural resources (minerals, metals, oil, coal, water, trees, parks) or producing/manufacturing products (livestock, lumber, automobiles, computers, etc.).

Assign each student two cities or towns. Have them put the names of their towns/cities on the class outline map at their correct location. Then ask them to gather information about their cities/towns such as products produced/manufactured or special features that contribute to state growth.

Students can begin their research with any of the magazines or newspapers that have articles on the two towns or cities. Then for further information students can write to local chambers of commerce or state and city governments. They can consult the library or ask friends and relatives who might live in the town/city to get information for them. Ideally, students should try to obtain photographs, brochures, posters, postcards, or significant items of the town/state that can be displayed.

As students receive responses to their letters and research, have them display them at the table and draw symbols of the information at the correct locations on the outline map. For example, if the city of Fairfax is known for its shoe manufacturing, the student should draw a picture of a shoe on the outline map at the Fairfax location.

When the map has been completed and the display table is all arranged, invite other classes to view it, allowing your students to serve as tour guides to the state fair table.

Follow-up: Based on the product information shown on the map, students can construct a "grown in our state" menu featuring foods grown or processed in the state. The menu can be designed in the shape of the state. With the help of parents, the students can prepare many of the food dishes and sponsor a food-tasting party featuring the foods from the state.

Activity 6-3-3 *Advertising our state*

Purpose: To identify features of a state that make it an attractive place in which to live

Materials:

1. drawing paper, markers
2. scissors, magazines

Procedure: Tell students that they are going to create advertisements showing why people would like to live in their state. More advanced students might design an ad that shows why their state is a good location for business.

The ads should reflect what the students know about the geography, climate, city life, and special opportunities of their state. Each ad should include a catchy slogan and an outline map containing drawings or magazine pictures of state features. When students have completed their ads, display them and analyze results.

Follow-up: Ask each student to interview a senior resident of the community, asking him/her the following questions:

- When and why did you come to the state?
- How has the state changed since you were born or since you moved here?
- What would you tell others to convince them to come to your state?

After all the interviews are done, have students share their information. How does it compare with the information students provided in their advertisements?

Additional Activities

• *Constructing relief maps*

Have students construct a three-dimensional relief map of their state, using one of the moldable mixtures described in Part Two of the handbook. The map can be painted the standard colors that indicate elevation levels.

• *From farm to city*

On a state map, locate three types of living environments: rural, suburban, and urban. Ask students to discuss how each of these areas is different. Encourage them to draw pictures or cut pictures from magazines to illustrate the different life-styles. More motivated students can construct a three-dimensional model of the three environments.

- *Analyzing local newspapers*

Try to obtain a selection of local newspapers statewide. Ask students to compare the news and local interest stories that are featured in the newspapers from each town or city. Have them discuss: Can differences in coverage be linked to differences in geography or economic activity?

As a class, create a state map with the names of all the towns or cities represented by the newspapers the students have gathered. Post the map, and attach articles or headlines from the various state newspapers, informing the readers of current news events.

Using Maps to Learn about U.S. and World Geography

Maps are the essential tools of the geographer. Thus, teaching geography without integrating map study is virtually impossible.

No doubt, you already have a large repertoire of map skills activities in your collection of teaching materials. The following ideas, however, provide some additional options to enliven instruction.

Activity 1-4-3 *The language of maps*

Purpose: To use geographic terms when describing an imaginary island

Materials:

1. large sheets of drawing paper
2. markers or crayons
3. copies of Supplement 1-4-3
4. dictionaries
5. a wall map

Procedure: (The following activity would be effective at the end of a unit on landforms.) Tell students that they have each discovered a brand new island, heretofore unknown to the world. As discoverers, they can name the island and all the natural features on it.

Have the students draw their islands and indicate the location and names of all the special features (mountains, volcanoes, rivers, lakes, deserts, forests, jungles, gulfs, bays, etc.) found there. Make sure students also draw in fictitious latitude/longitude coordinates. You may want to use Supplements 1-4-3 a, b, c to provide students with ideas of the various geographic features that can be included on their island.

Encourage students to choose thematic names for their islands, to which the names of other features could be linked. For example, President Island might have a Polk Peninsula, Roosevelt River, Reagan Ravine, Bush Bay, Carter Cape, and so on. The sketch below provides another example.

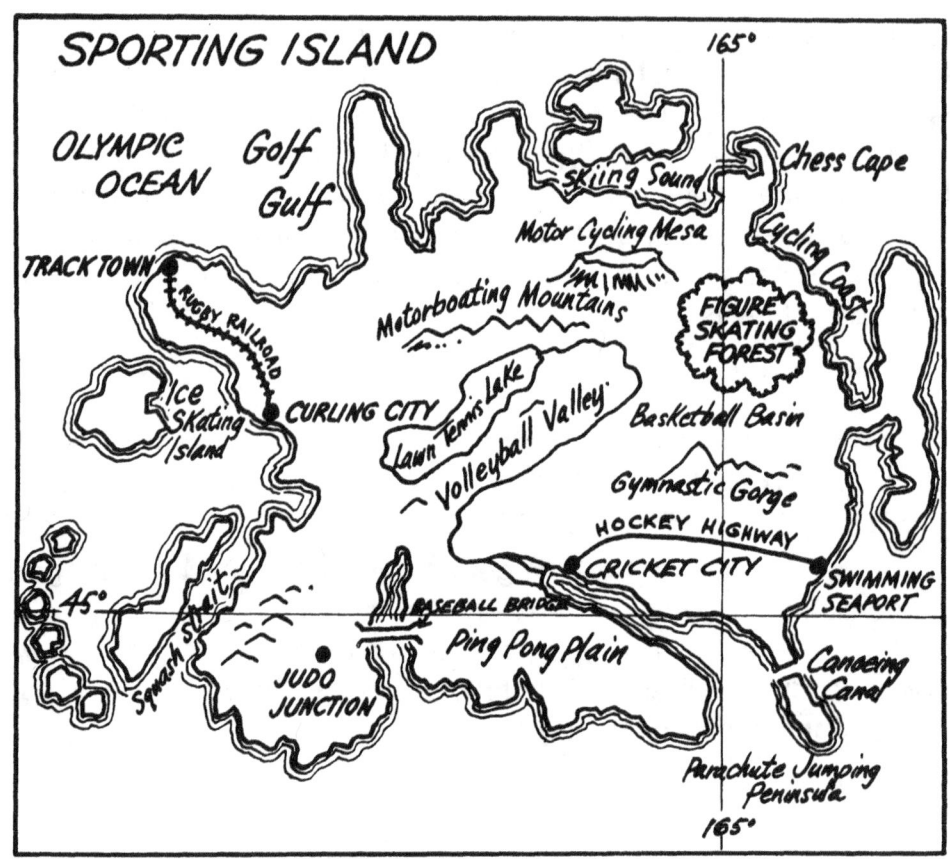

Follow-up: Some students might like to construct three-dimensional models of their islands. If the models are constructed in a pan as shown below, water can be added to make the models seem more realistic.

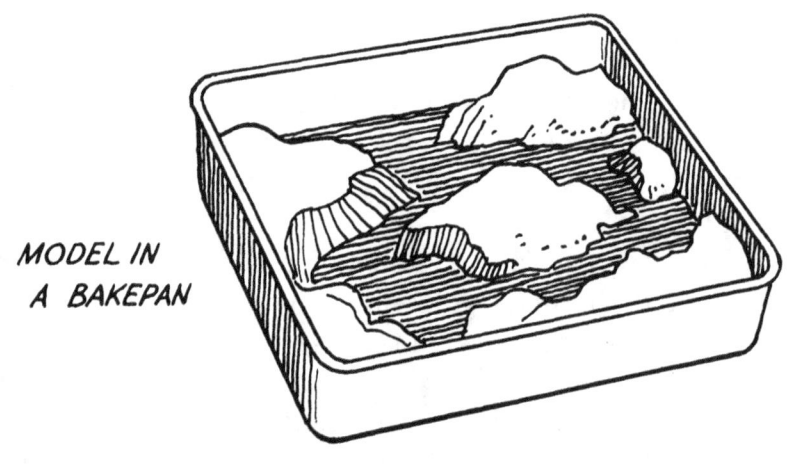

MODEL IN
A BAKEPAN

Activity 2-4-3 *Regions of the United States*

Purpose: To locate regions of the United States on a map and name products for which each region is known

Materials:

1. copies of a political map of the U.S. from the Map Masters section of the appendix
2. a wall map showing physical features of the United States
3. student copies of Supplement 2-4-3

Procedure: Write the following headlines on the chalkboard:

- Earthquakes Shake Pacific Coast States
- Great Lakes States Sending Acid Rain to New England
- People Moving to Southwest at Record Rate
- Hurricane Hits Southeast
- Rock Group to Tour Mid-Atlantic States

Ask the students to tell you the location of the events mentioned in the headlines (regions of the United States). Explain that the United States can be divided into regions according to climate, landforms, economic activities, etc.

Distribute Supplement 2-4-3. Point out that in this particular case the United States was divided into regions based on geography, although the states in each region have some cultural and economic similarities.

Give students copies of a U.S. outline map and have them mark the regions. Then have them compare their regional maps with the physical map of the United States to analyze how each region is geographically distinct from its neighbors.

Divide the class into seven groups, assigning each group one of the regions. Over the next several days, have students research the economic activities of their region. They should list the important products of the region and draw them on their maps.

Follow-up: For a week, have students collect newspaper articles and make notes of television news reports that refer to specific regions of the United States. After a week, have them share the results.

Activity 3-4-3 *Industrial areas of the United States*

Purpose: To identify the different types of industrial areas in the United States

Materials:

1. large sheets of newsprint, tagboard, or butcher paper
2. yardsticks, marking pens, pencils
3. an outline map of the United States
4. magazines

Procedure: Tell students that you want them to make a map of the United States showing the locations of the greatest concentration of jobs. Review the procedures for using grids, presented in Part Two of the handbook. Explain how we can use grids to enlarge maps.

Give each student an outline map of the United States. Direct them to mark off one-inch squares over their maps and give an alphabetical (vertical) and numerical (horizontal) value to each block of the grid. Next, have students repeat this procedure on a large piece of paper. This time, however, they should make four- or five-inch squares.

Using each of the squares from the smaller map as references, have students draw a giant outline of the U.S. map, including state boundaries. Tell them to draw it first with a pencil and later trace over the lines with a marker.

When students have enlarged their maps, divide the students into groups of four. Have the members of each group select the best map from their group and divide it into four sections. Assign one of the sections to each of the students in the group. Each will be responsible for finding out the largest industries in the states that are located in their section of the map. Here are some examples:

- Michigan: autoworkers/automobile industry
- Louisiana: offshore oil workers/petroleum industry
- Kansas: wheat farmers/agricultural industry
- New York: financiers/banking industry
- California: film production technicians/movie industry
- Florida: hotel managers/tourist industry

Students can use special products or occupation maps in atlases to research their section, or they can look up the individual states in the encyclopedia. Make sure the students understand that there are a range of jobs in every location but that some industries are concentrated in particular areas.

Once the research is completed, have students do one of the following:

- write their results on their group map
- draw symbols to represent the occupations/industries in their area
- attach magazine pictures that show people engaged in jobs that are common in their area

Follow-up: Interested students might like to use the above activity to explore one of the seven regions of the United States (as in Supplement 2-4-3). After making a giant outline map of a selected region (see illustration on page 77), students can display products, or symbols of products and services, for which the region is known.

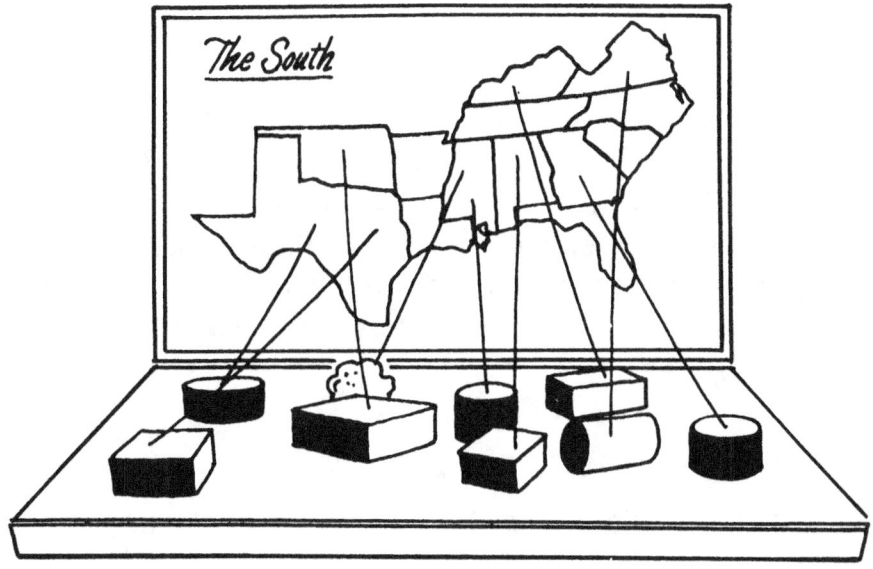

Activity 4-4-3   *Tracing transportation routes*

Purpose: To know how to follow a route on an interstate highway map and indicate the directions you take toward a destination

Materials:

1. copies of Supplement 4-4-3
2. marking pens

Procedure: Explain the relationship between transportation routes and the growth of economic regions. (Transportation makes it possible for products to get from where they are produced to where people need them.)

   Pass out copies of Supplement 4-4-3 and ask students to complete the exercise.

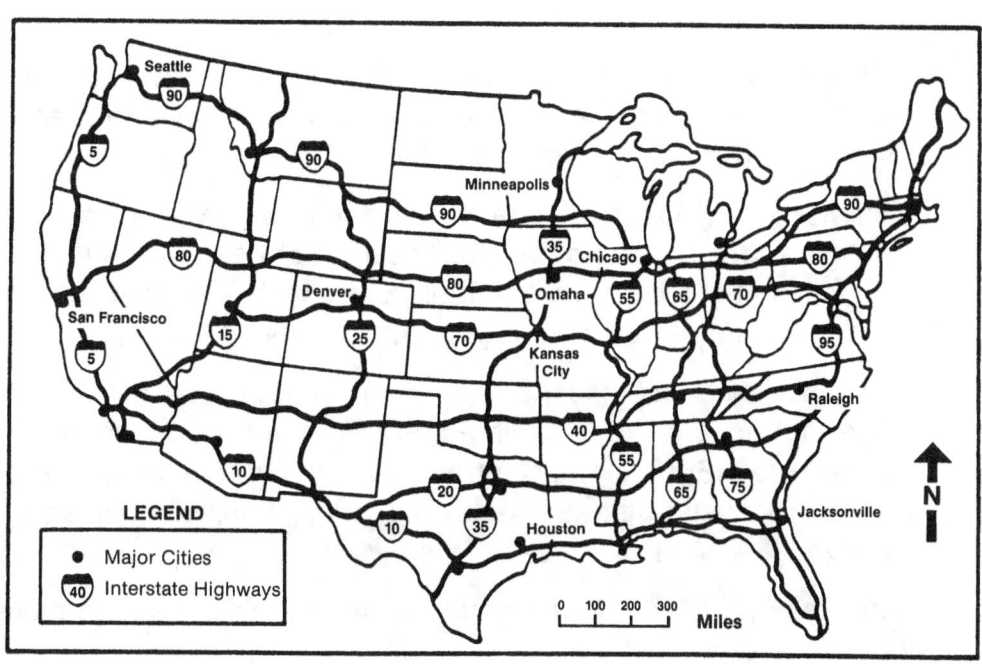

Follow-up: As additional reinforcement, have students design an activity that is based on a sea route that a captain of a cargo ship might take while following orders to pick up and deliver cargo from ten different ports. Students can use a world map and cardinal directions to determine their route. After students have planned their itinerary sheets, have them exchange papers and complete the information on the sheets.

Activity 5-4-3 *Clothing, shelter, and animals from around the world*

Purpose: To create thematic maps based on types of clothing, shelter, and animals

Materials:

1. three large world outline maps
2. reference materials

Procedure: Review the definition of a thematic map (a map that provides only one kind of information). Divide the class into three groups and assign each group one of the following topics:

- clothing from around the world
- shelters from around the world
- animals from around the world

Have the students research their topics and place the information in two columns: *country and item* for cultural items, or *continent and animal* for animals. You may need to provide the students with a few of the following examples:

Clothing: turban (India), kilt (Scotland), parka (Arctic), kimono (Japan), sombrero (Mexico), muumuu (Hawaii), poncho (Peru), sarong (Java), fez (Turkey), sari (India), bolero (Spain)

Shelter: yurt (Mongolia), chalet (Switzerland), thatched-roof huts (Central Africa), stilt homes (Indonesia), adobe (southwestern United States), flat-topped homes (Middle East), tents (North Africa), stone houses (Central Europe), houseboats (China)

Animals: seal, walrus, caribou (Arctic); bear, buffalo, coyote, deer (North America); anteater, llama (South America); panda, yak, tiger, horse, boar (Europe and Central Asia); elephant, giraffe, zebra, lion, gorilla (Africa); fruit bat, kangaroo, dingo, koala, wombat, and platypus (Australia)

After students have gathered information on their topics, have them convert it into a thematic map.

Follow-up: Post pictures of such world-famous buildings and structures as the following:

- Pyramids (Egypt)
- Eiffel Tower (France)
- Taj Mahal (India)
- Great Wall of China (China)
- White House (United States)
- Tower of London (Great Britain)
- Kremlin (Soviet Union)
- Vatican (Italy)
- United Nations (United States)
- Wailing Wall (Israel)
- the Great Mosque al Haram (Saudi Arabia)

Have students circle these locations on a world desk map or print the name of the building at its location in the country.

Additional Activities

- *Geographic features flash cards*

Give each student the names of one or more geographic features listed on Supplement 1-4-3. For each term, students are to prepare a 5" x 8" file card that shows an illustration of the feature and gives its definition. The completed cards can be put into a learning center for future reference, or the terms can be alphabetized with their definitions and converted into a glossary.

- *Origins of geographic terms and place names*

Pass out Supplement 6-4-3 and discuss the samples given. Ask the students to complete the worksheet.

- *Building a cross section of a U.S. relief map*

Based on the sketch below, have students construct a portion of a U.S. relief map that shows a cross section. Ask them to label the features or key them to an accompanying relief map of the United States.

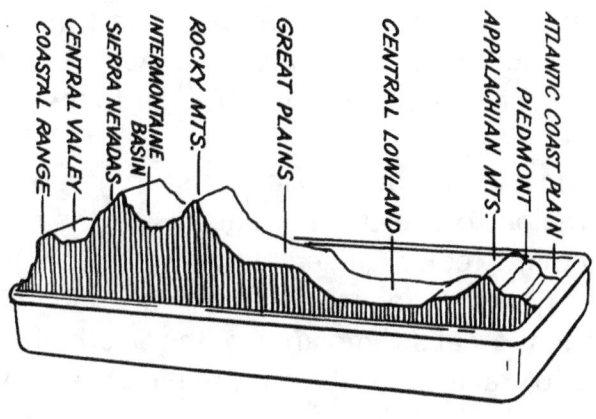

• *Global economics*

After the class has studied the products of various countries or regions, ask each student to create a practice worksheet for classmates. The worksheets should list 20 countries, their latitude/longitude coordinates, and major products; the three lists should be randomly ordered so that students will have to match the name with the coordinates and the major product.

Using Maps to Learn about U.S. History

Although geography directly influences the course of history, many textbooks do not adequately develop this causal relationship. If you have not previously emphasized the relationship between history and geography, the activities in this section will help you present lessons based on this causal relationship.

Activity 1-5-3 *Mapping the location of American Indian tribes*

Purpose: To identify the areas where Americans lived prior to European colonization of North America

Materials:

1. large outline map of the United States (can be an enlarged copy of the physical map located in the Map Masters section of the appendix)
2. 3" x 5" cards
3. colored pencils
4. reference materials on American Indians

Before beginning this activity, ask students to label the eight major regions of American Indian tribes on their U.S. desk outline maps: the Northeast, the Southeast, the Great Plains, the Southwest, the Great Basin, California, the Plateau, and the Northwest. Have them use a different color pencil to lightly color in each region.

Procedure: Have students use reference books to help them locate where Indian tribes were living prior to European colonization. Ask them to write in the names of the tribes in the prospective states. You can refer to the map on page 81 if students need help.

Divide the class into eight groups, assigning each group responsibility for one of the eight Indian regions of the United States. Have them give examples of how the geography of the region influenced the manner in which the Indians lived. That is, list the type of food, housing, clothing, and transportation that resulted from living in the wet swamplands of the South, the dry rocky deserts of the West, or in the cool, rainy coastal region of the great Northwest.

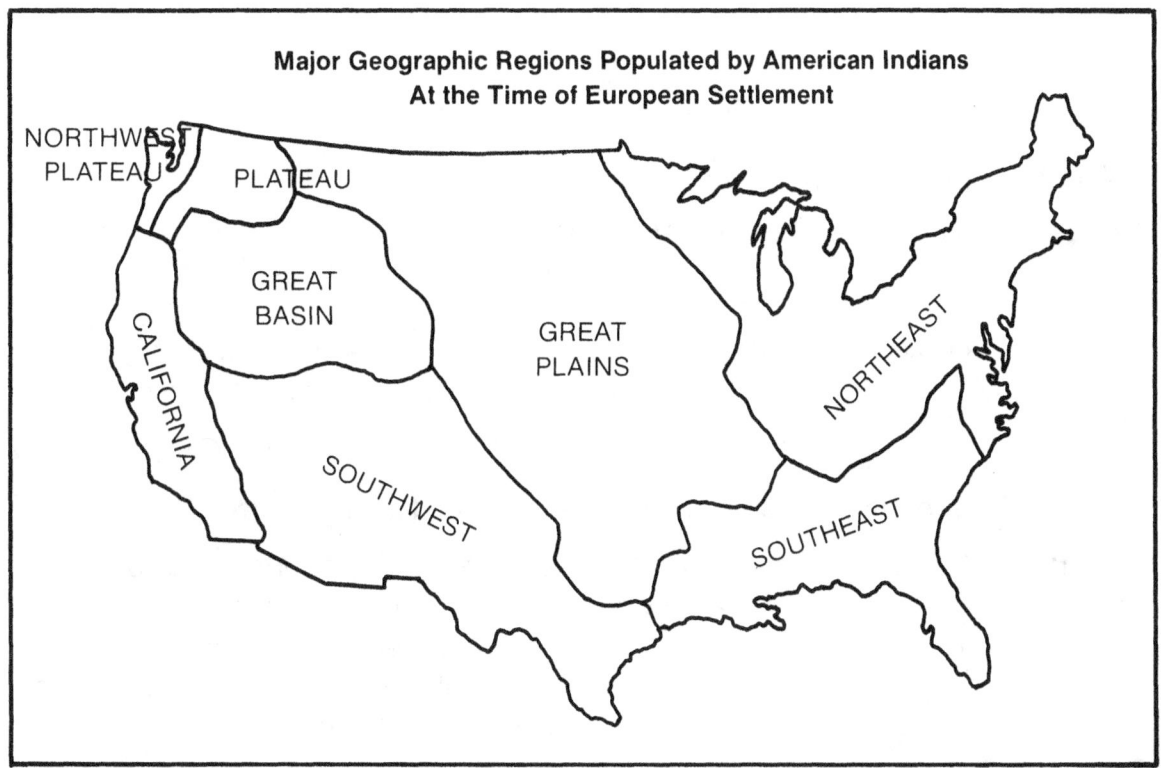

**Major Geographic Regions Populated by American Indians
At the Time of European Settlement**

Follow-up: Point out that various American Indian tribes from the Southeast and the Northwest were relocated to Indian Territory (Oklahoma) in the nineteenth century. Talk about the different environments from which these tribes came, and explain how the differences affected the way in which they had to live in the new land.

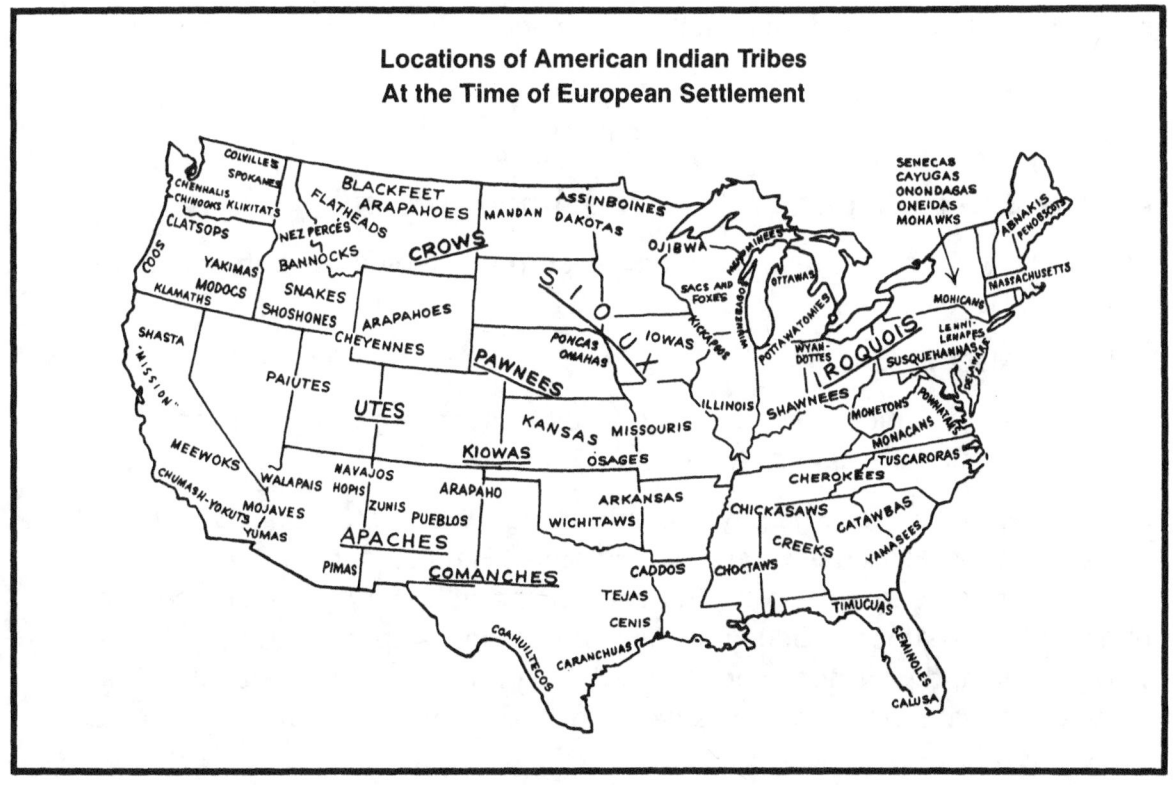

**Locations of American Indian Tribes
At the Time of European Settlement**

Activity 2-5-3 *European settlement in North America*

Purpose: To trace the European origins of many American place names

Materials:

1. copies of the maps of North America, Europe, and the United States from the Map Masters section of the appendix
2. markers or crayons
3. reference materials such as encyclopedias and atlases

Procedure: Distribute copies of an "Old World" and "New World" map to students. Ask them to choose a European country and a region of the United States or Canada that has the same or similar name (e.g., New Brunswick). Have students use reference materials to locate the origin of the names. They should place the names on their outline map and color-code it with the color of its national origin.

When students have completed the maps, help them draw conclusions about settlement patterns in the New World. Did European settlers tend to migrate to areas that had climates or landforms similar to those in their home countries?

Follow-up: Students could conduct additional research to determine whether or not their conclusions about settlement patterns were accurate. Other students could research cultural contributions from each of the European countries studied, placing pictures representing these contributions around the two maps.

Activity 3-5-3 *Geography and the American Revolution*

Purpose: To cite three ways in which geography affected the outcome of the American Revolution

Materials:

1. multiple copies of student atlases or large physical and climate maps of the United States

Procedure: Display a large physical and climate map of the United States. Show the location of the thirteen American colonies. Help students understand the important part geography plays in the outcome of history. For instance, the harsh winter weather of December 1776 could have wiped out George Washington and his troops and changed the course of history, had it not been for the fact that European armies were unaccustomed to fighting during the winter months.

Besides weather, difficult terrain and the absence of water or food supplies also slowed down marching troops and affected battle strategies. In addition, British troops were unaccustomed to the American countryside and came unprepared for surprise attacks and for a type of warfare that was fought behind trees and hills.

England's Thirteen Colonies

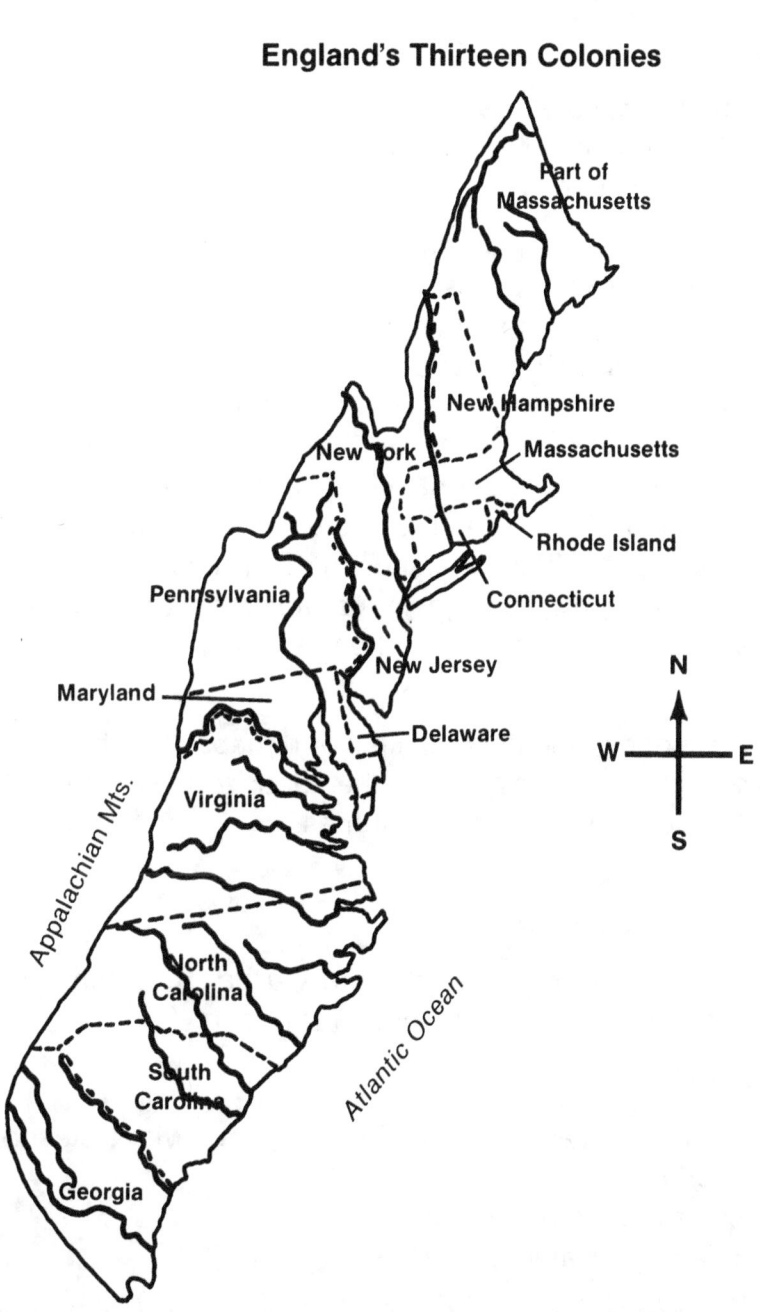

Follow-up: Encourage interested students to research the influence of geography on other wars in which the United States has been involved.

Activity 4-5-3 *Westward expansion*

Purpose: To correlate a map with a timeline to show settlement patterns and the development of the American West

Materials:

1. outline maps of the United States
2. rulers, crayons or marking pens
3. reference books

Procedure: Using reference materials (including history textbooks), have students create expansion maps, showing the sites of early forts and settlements in the West, river portages, mountain passes, routes and trails, territorial acquisitions, dates of statehood, and so on.

Beneath their maps, students should create a timeline covering the period from 1787 (Northwest Ordinance) to 1912 (admission of Arizona). The dates of significant events shown on their maps should be added to the timeline in chronological order. Remind students that timelines, like maps, should be drawn to scale. If necessary, provide assistance in determining a scale for their timelines. Students could color-code major territorial acquisitions with entries on the timeline or devise other such schemes for correlating their maps and timelines.

Follow-up: Challenge students to make world maps that show U.S. territorial acquisitions outside the boundaries of the continental United States during the late nineteenth and early twentieth centuries.

Activity 5-5-3 *Urbanization*

Purpose: To locate major nineteenth-century urban areas on a map and make hypotheses about reasons for their locations

Materials:

1. large physical map of the United States with a clear plastic overlay
2. color markers
3. reference materials

You might want to write the following list of cities on the chalkboard prior to the class period or photocopy them and distribute them to the students:

New York	Boston	Washington
Philadelphia	Cincinnati	Newark
St. Louis	New Orleans	Louisville
Chicago	San Francisco	
Baltimore	Buffalo	

Procedure: Review the reasons for the growth of cities in the 1800s. Draw students' attention to the list of cities on the chalkboard. All of these cities experienced significant growth in the 1800s; by 1870, each of these cities had more than 100,000 inhabitants.

Using a political map as a reference, have students mark the locations of these cities on the overlay. Ask the students what they notice most about these cities. (They are located along oceans, lakes, or rivers.) Why was this so important? (Settlers came to the cities on water transportation; goods were shipped into and out of the cities by water.)

In the west, cities grew up along another form of transportation. Ask students which form of transportation this was. (railroads) Using reference materials, have students locate the routes of the following railroads and add them to the overlay:

Northern Pacific	Union/Central Pacific
Southern Pacific	Atchison, Topeka, and Santa Fe

Have students speculate on the difficulties encountered in building these railroads.

Follow-up: Encourage students to find out which states were most and least urbanized in 1900, and create a map showing the results of their research.

Additional Activities

• *Influence of rivers, past and present*

Have individuals or groups of students research one of the following great river systems of the world:

Nile	Amazon	Volga	Mississippi
Rhine	Ganges	Yangtze	Murray-Darling

Be sure students gather information on the river's flow pattern, tributaries, relief factors, and cultural features (cities and their locations) as well as on historical events that took place along the river.

Each group should prepare a model of their river system and adjacent topographic and relief features. The models should be accompanied by written reports on the river's past and present importance to people.

- *Conserving our wilderness*

Point out that when settlers first came to the United States, the supply of land seemed endless. By the early 1900s, however, some people began to worry that all our natural resources would be destroyed as people settled the West.

Have students conduct research on the various categories of land (both state and national) that have been set aside for limited use by people. Some students may want to report on recent controversies regarding the use of public lands, using maps to illustrate their reports.

- *The dust bowl*

Explain how the creation of a dust bowl in the Great Plains region of the United States in the 1920s and 1930s showed the combined effects of humans and nature. Help students simulate the creation of the dust bowl. Bring two squares of sod into the classroom. Allow both to dry out. In one allow the plants to remain untouched; in the other, destroy the root systems of the plants (simulating plowing). After the two squares have thoroughly dried, blow a fan across both. Which is more susceptible to wind? Why?

Have students research the conditions that created the dust bowl. They should determine from which direction the prevailing winds in the dust bowl blew. Challenge students to find out how far east the effects of the dust bowl were felt in the form of dirt storms.

Using Maps to Learn about Current World Events

Our nation is inextricably bound to other countries around the world. One way of helping students to understand these relationships is to provide them with frequent opportunities to focus on current world events and discuss how they affect their lives. The activities in this section will provide such opportunities.

Activity 1-6-3   *Weekly headline maps*

Purpose: To summarize current event stories and indicate on a world map where the events are occurring

Materials:

1. outline map of the world from the Map Masters section of the appendix
2. overhead projector
3. butcher paper or newsprint
4. color markers, note cards, pushpins

Procedure: Using the overhead projector and outline map, make a large outline map of the world on butcher paper or newsprint. Ask each student to bring in one news item per week and summarize the story on a note card; different colored note cards could be used to represent the various continents, if desired. When students have completed their summaries, they should attach the headlines to the large world map and post the cards around the map.

At the end of each week, discuss all the stories that have been posted, focusing especially on the evidence of interdependence among nations. Before the next week's stories come in, take down the old summary cards and put them in a "morgue" for future reference.

Follow-up: When students have done this exercise several times, ask them to rank the top ten stories of the week. Mark the stories with a color pushpin or flag.

At the beginning of each week you may try putting question marks at locations on a world map where news stories are likely to occur. Challenge students to find out what is happening in those locations.

Activity 2-6-3 *Geography and world events*

Purpose: To make inferences about the relationship between geography and world events

> **Materials:**
>
> 1. U.S. and world maps covered with heavy-duty acetate
> 2. grease pencils or water-soluble markers
>
> The maps can be large wall maps or spring-roller mounted maps. If you use spring-roller maps, attach the acetate to the edges of the maps.

Procedure: Assign student reporters to find one or two events in the newspaper each day. They should write key words related to the events at the correct location on the map overlays.

Spend five minutes each day examining the posted information and helping students look for relationships between geography and the events. These relationships may be as simple as the Alaskan oil spill, occurring near the end of the Alaskan Pipeline, or they may be much more complex, involving the wind patterns that carry acid rain from one area to another.

Follow-up: Provide students with outline maps of the continents from the Map Masters section of the appendix. You may give certain students the responsibility of becoming experts in particular continents, or rotate responsibility for the continents from week to week. In either case, students should post the weekly events on transparencies and project them on an overhead projector for class discussion.

Activity 3-6-3   *Connecting global current events*

Purpose: To draw a network of international relationships

Materials:

1. a large outline map of the world
2. color markers
3. variety of news articles on international events

Procedure: Divide the class into small groups, giving each group several articles to read. Ask students to make a list of all the countries directly involved in or affected by each event. Using a separate color marker for each event, students should draw lines connecting all the countries involved in that particular event.

When all groups have added their information to the map, discuss the results. How many areas is the United States linked to? What are the implications of these linkages? What other countries or regions have many international linkages?

Follow-up: Encourage students to expand the global network on their world maps as they keep abreast of world affairs.

Additional Activities

● *Grid quiz*

Superimpose an alphanumeric grid system over a blank outline map of the United States or world. Have students identify current events that occurred in specific squares and write questions concerning those events. For example, a question might be: "Write an A in the square where the vice-president gave a speech on Central America." Questions can be written so that, when all the answers are written on the grid, the letters spell out a current events word.

● *News sphere*

Draw a large circle on newsprint or butcher paper, putting your community in the center and showing the cardinal directions to the sides. Point out to students that the initials of the cardinal directions spell out "NEWS."

Students are to use this news sphere to indicate where events in state and national news are happening: north, south, east, or west of your community. Headlines or news stories can be mounted on the sphere in the correct quadrant. An example is shown on page 89.

NEWS SPHERE

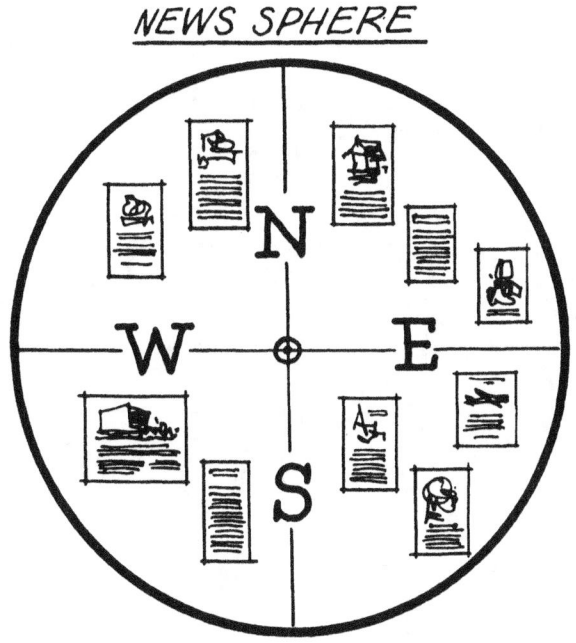

● *Current events contest*

Every week, reproduce an outline map with 10 to 20 numbered locations. Each number should represent the site of a significant event that occurred during the past week. You may provide an accompanying list of statements summarizing the events that students can match with the map, or you can challenge students to come up with the events on their own. This activity can be done individually or in groups. The first group or student to identify all the locations is designated as the weekly winner.

Part FOUR

Student Supplements

Part FOUR

Student Supplements

The supplements that follow are designed to work with the specified lessons. You may wish to adapt particular supplements to a given grade level or use a supplement as a starting point for an exercise that you write.

Symbols and Signs 93

Current Members of the United Nations 96

Highway Map Symbols 98

Suggested Symbols for Sketch Mapping 99

Scrambled States 103

Compass Roses and Other Tools 107

Tracing Routes 109

Hidden Words 112

Map Search 113

What's in a Name? 116

Cities Having the Same Name 119

What Is Scale? 122

Different Kinds of Scales 124

Using Scale 125

A Grid 126

Location Coordinates and the World Map 128

Coordinates and Capitals 131

What's on the Menu? 136

Making a Flight Plan 139

Tallest Buildings in the United States 140

True or False? 143

Symbols for Weather Elements and Cloud Types 144

Symbols for Barometric Tendency 145

Beaufort Scale of Winds 146

Weather Report from One Station 147

Making Inferences 148

Making Simple Floor Plans 149

Safety Symbols 150

Stop, Look, and Listen When You Travel 152

Place Names 153

Natural Geographic Features 157

Regions of the United States 160

Planning a Highway Route 161

Finding the Origin of Geographic Terms and Place Names 163

Symbols and Signs

Learning to read road signs that do not use words will help you receive information quickly. The messages below appear as pictures on road signs. Can you match each of the messages on page 94 with one of the symbols below? Put the number of the symbol next to the message it represents.

Symbols and Signs

_____ slippery area

_____ DO NOT ENTER

_____ buses permitted

_____ drinking fountain

_____ first aid station

_____ no bicycles permitted

_____ two-way traffic

_____ campground

_____ women's rest room

_____ taxi station

_____ no smoking permitted

_____ information available

_____ cars permitted

_____ farm traffic ahead

_____ turn right

_____ airport

_____ restaurant

_____ no cars permitted

_____ men's rest room

_____ rest rooms (women & men)

_____ dangerous curve

_____ handicapped persons

_____ smoking permitted

_____ no buses permitted

_____ telephones available

_____ bicycles permitted

_____ no dogs allowed

_____ taxi and bus station

_____ roadside park

_____ hotel-motel available

Supplement 3-1-2 (answer key)

Symbols and Signs

28	slippery area	21	airport
10	DO NOT ENTER	20	restaurant
3	buses permitted	2	no cars permitted
19	drinking fountain	16	men's rest room
18	first aid station	17	rest rooms (women & men)
8	no bicycles permitted	9	dangerous curve
30	two-way traffic	27	handicapped persons
24	campground	13	smoking permitted
15	women's rest room	4	no buses permitted
5	taxi station	25	telephones available
14	no smoking permitted	7	bicycles permitted
26	information available	22	no dogs allowed
1	cars permitted	6	taxi and bus station
29	farm traffic ahead	11	roadside park
12	turn right	23	hotel-motel available

Supplement 4-1-2

Current Members of the United Nations

Afghanistan
Albania
Algeria
Angola
Antigua and Barbuda
Argentina
Australia
Austria
Bahamas
Bahrain
Bangladesh
Barbados
Belgium
Belize
Benin
Bhutan
Bolivia
Botswana
Brazil
Brunei
Bulgaria
Burkina Faso
Burma
Burundi
Byelorussia
Cameroon
Canada
Cape Verde
Central African Republic
Chad
Chile
China
Colombia
Comoros
Congo
Costa Rica
Cuba
Cyprus
Czechoslovakia
Denmark

Djibouti
Dominica
Dominican Republic
Ecuador
Egypt
El Salvador
Equatorial Guinea
Ethiopia
Fiji
Finland
France
Gabon
Gambia
Germany, East
Germany, West
Ghana
Greece
Grenada
Guatemala
Guinea
Guinea-Bissau
Guyana
Haiti
Honduras
Hungary
Iceland
India
Indonesia
Iran
Iraq
Ireland
Israel
Italy
Ivory Coast
Jamaica
Japan
Jordan
Kampuchea (Cambodia)
Kenya
Kuwait

Laos
Lebanon
Lesotho
Liberia
Libya
Luxembourg
Madagascar (Malagasy)
Malawi
Malaysia
Maldives
Mali
Malta
Mauritania
Mauritius
Mexico
Mongolia
Morocco
Mozambique
Nepal
Netherlands
New Zealand
Nicaragua
Niger
Nigeria
Norway
Oman
Pakistan
Panama
Papua New Guinea
Paraguay
Peru
Philippines
Poland
Portugal
Qatar
Romania
Rwanda
Saint Christopher and
  Nevis
Saint Lucia

Current Members of the United Nations

Saint Vincent and the Grenadines	Sri Lanka	USSR (Soviet Union)
Samoa (Western)	Sudan	United Arab Emirates
Sao Tome and Principe	Suriname	United Kingdom
Saudi Arabia	Swaziland	United States
Senegal	Sweden	Uruguay
Seychelles	Syria	Vanuatu
Sierra Leone	Tanzania	Venezuela
Singapore	Thailand	Vietnam
Solomon Islands	Togo	Yemen
Somalia	Trinidad and Tobago	Yemen, South
South Africa (suspended)	Tunisia	Yugoslavia
Spain	Turkey	Zaire
	Uganda	Zambia
	Ukraine	Zimbabwe

HIGHWAY MAP SYMBOLS

ROAD CLASSIFICATIONS

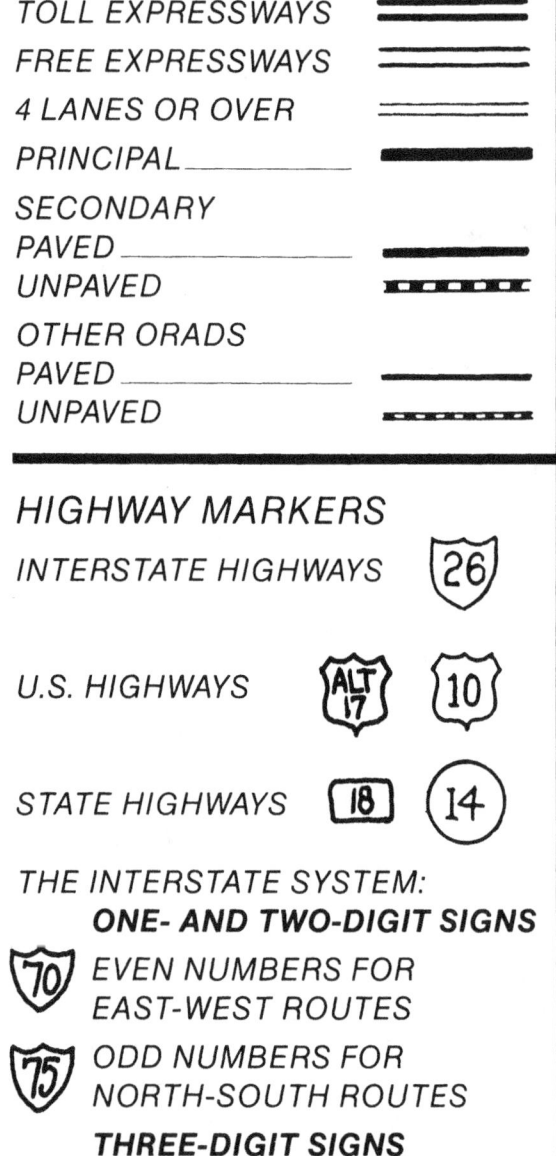

TOLL EXPRESSWAYS

FREE EXPRESSWAYS

4 LANES OR OVER

PRINCIPAL

SECONDARY
PAVED

UNPAVED

OTHER ORADS
PAVED

UNPAVED

HIGHWAY MARKERS

INTERSTATE HIGHWAYS 〔26〕

U.S. HIGHWAYS 〔ALT 17〕 〔10〕

STATE HIGHWAYS 〔18〕 〔14〕

THE INTERSTATE SYSTEM:
ONE- AND TWO-DIGIT SIGNS

〔70〕 EVEN NUMBERS FOR
EAST-WEST ROUTES

〔75〕 ODD NUMBERS FOR
NORTH-SOUTH ROUTES

THREE-DIGIT SIGNS

〔265〕 ROUTES AROUND OR THRU-
CITIES. FIRST DIGIT EVEN.

〔195〕 SPUR INTO A CITY.
FIRST DIGIT ODD.

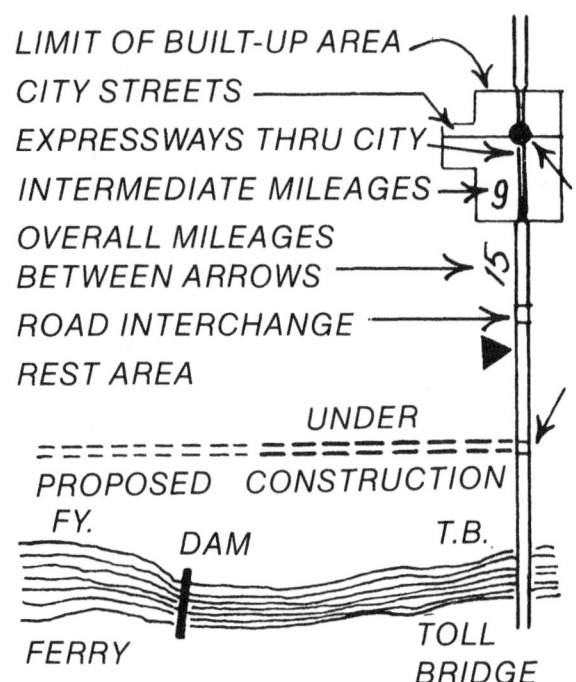

LIMIT OF BUILT-UP AREA

CITY STREETS

EXPRESSWAYS THRU CITY

INTERMEDIATE MILEAGES 9

OVERALL MILEAGES
BETWEEN ARROWS 15

ROAD INTERCHANGE

REST AREA

UNDER
PROPOSED CONSTRUCTION

FY.
DAM T.B.

FERRY TOLL
BRIDGE

POPULATIONS OF CITIES AND TOWNS

● 0—10,000

○ 10,000—25,000

◉ 25,000—100,000

□ 100,000—250,000

▣ 250,000—1,000,000

■ over 1,000,000

SUGGESTED SYMBOLS FOR SKETCH MAPPING

RAILROAD	CHURCH	FERRY
BUILDINGS	CEMETERY	POWER TRANSMISSION
FARM BUILDINGS	CANAL	**BOUNDARIES** INTERNATIONAL — ·· — ·· — STATE — · — · — COUNTY ——— MILITARY STATION
TUNNELS	ORCHARD	SCENIC ROUTES HIKING TRAIL
AIRPORT HELIPORT	PIPELINE	RIVER and STREAM
BRIDGE	DAM	HOSPITAL

Supplement 5-1-2 (c)

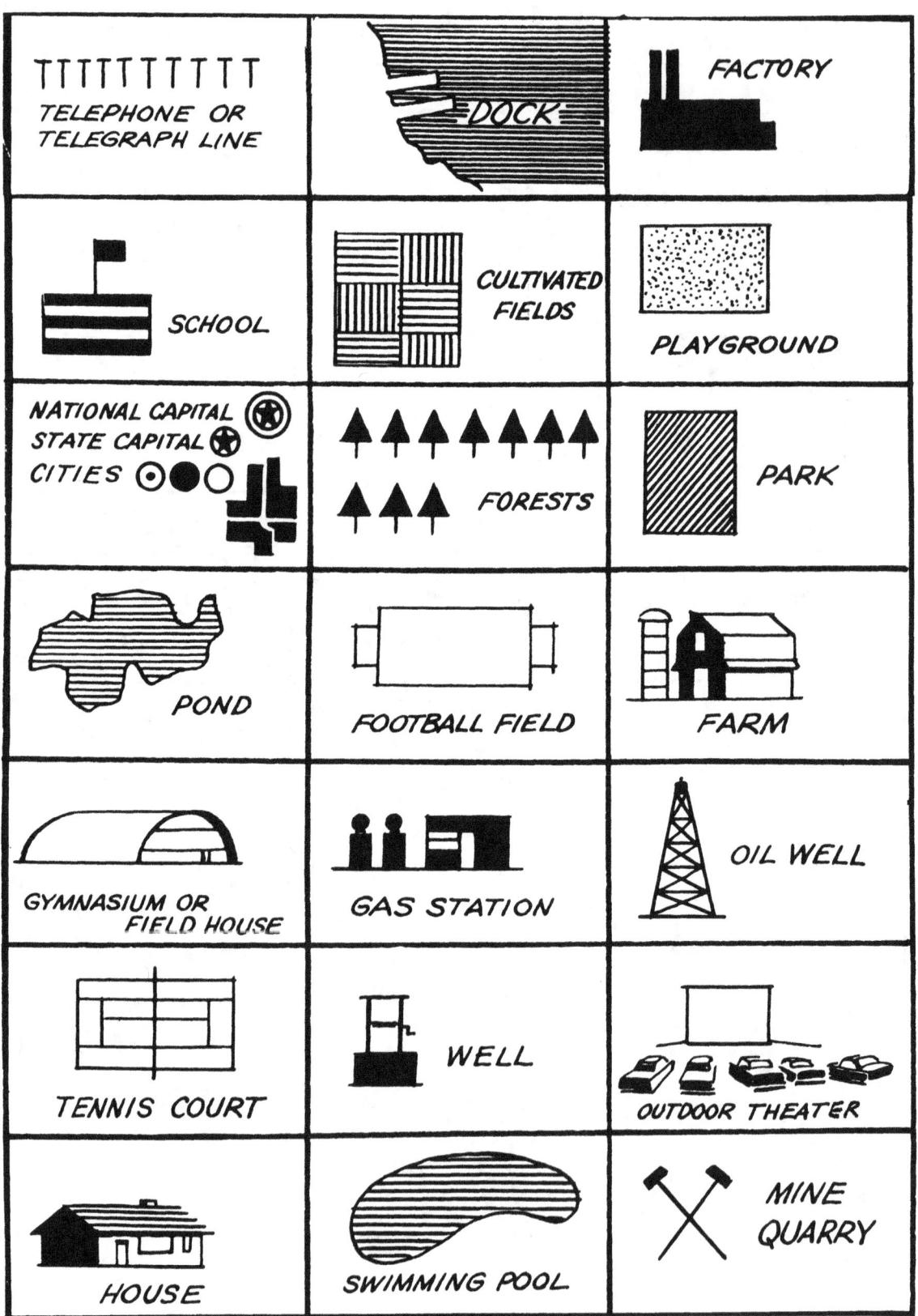

TELEPHONE OR TELEGRAPH LINE

DOCK

FACTORY

SCHOOL

CULTIVATED FIELDS

PLAYGROUND

NATIONAL CAPITAL
STATE CAPITAL
CITIES

FORESTS

PARK

POND

FOOTBALL FIELD

FARM

GYMNASIUM OR FIELD HOUSE

GAS STATION

OIL WELL

TENNIS COURT

WELL

OUTDOOR THEATER

HOUSE

SWIMMING POOL

MINE QUARRY

100

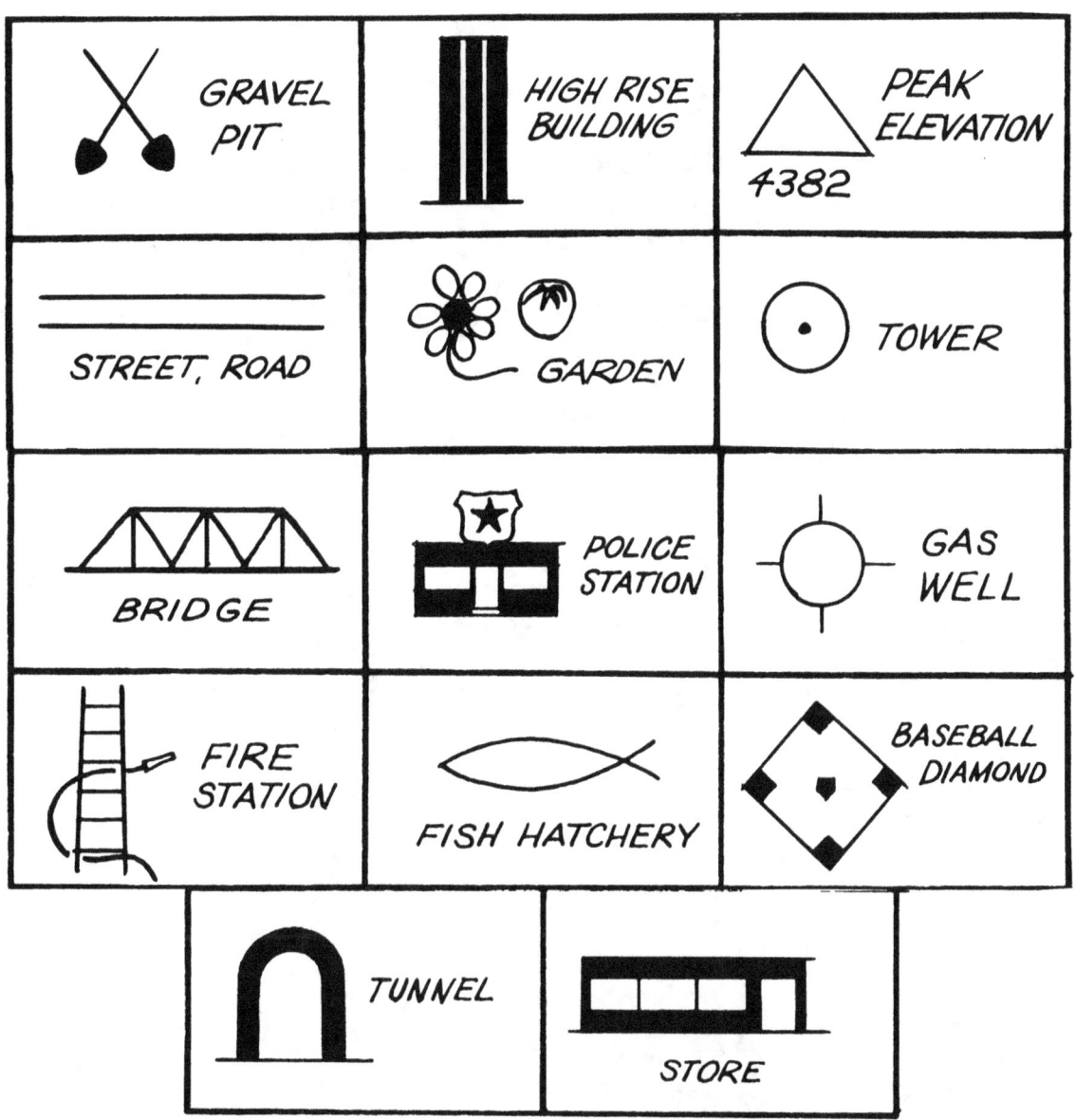

GRAVEL PIT

HIGH RISE BUILDING

PEAK ELEVATION 4382

STREET, ROAD

GARDEN

TOWER

BRIDGE

POLICE STATION

GAS WELL

FIRE STATION

FISH HATCHERY

BASEBALL DIAMOND

TUNNEL

STORE

POINTS OF INTEREST

AIRPORTS
COMMERCIAL ——————
(AIRLINES)

COMMERCIAL (OTHERS)

MILITARY ——————

MONUMENTS

HISTORICAL —————

NATIONAL —————

PARKS

STATE ————————

NATIONAL————— ★

OTHER INFORMATION

COUNTY BOUNDARY — — — — —   POINT OF INTEREST— ■

STATE BOUNDARY — —·— —·—

WILDLIFE AREAS    SKI AREA ———————

FISH HATCHERIES    CAMP AREA ——————— Ⓒ

COVERED BRIDGE———    ELEVATION POINT ·———— △

Scrambled States

The list below gives the names of the fifty states with their letters scrambled. Unscramble each one and write the correct spelling in the left-hand column. Then choose the correct abbreviation for the state from the attached list. Write the abbreviation in the second column and also on the outline map of the United States.

State	Abbreviation	Scramble
1. _____	_____	ERVNOMT
2. _____	_____	TUHOS KATAOD
3. _____	_____	GINOAWTNHS
4. _____	_____	AIWO
5. _____	_____	TMAINSNOE
6. _____	_____	RNIIIGVA
7. _____	_____	DORCOOLA
8. _____	_____	IAGGROE
9. _____	_____	HRNTO RLANAICO
10. _____	_____	WEN SEJREY
11. _____	_____	IHOO
12. _____	_____	NARLYMDA
13. _____	_____	STWE RGIINIAV
14. _____	_____	IIOWSSCNN
15. _____	_____	KENARABS
16. _____	_____	SSSSPMPIIII
17. _____	_____	OHNRT OKDAAT
18. _____	_____	BLAAAAM

Scrambled States

State	Abbreviation	Scramble
19. _____	_____	SOINLILI
20. _____	_____	KKYUTNCE
21. _____	_____	HTAU
22. _____	_____	MIENA
23. _____	_____	RIAONZA
24. _____	_____	PVINNYSEALNA
25. _____	_____	STHAMSTSUSAEC
26. _____	_____	NNIIAAD
27. _____	_____	SSANRAKA
28. _____	_____	NAICHMIG
29. _____	_____	UIEONNCCTTC
30. _____	_____	NESNESETE
31. _____	_____	ROICIALANF
32. _____	_____	WEN ICXOEM
33. _____	_____	IHOAD
34. _____	_____	TSOHU NRLAACIO
35. _____	_____	OIFDALR
36. _____	_____	YMGINOW
37. _____	_____	NWE PHSEMHRIA
38. _____	_____	HEROD NLASID
39. _____	_____	OSAUAILNI
40. _____	_____	ENW OKYR

Scrambled States

State	Abbreviation	Scramble
41. _____	_____	OHAKLOAM
42. _____	_____	DEEAALRW
43. _____	_____	MONAANT
44. _____	_____	DAAVEN
45. _____	_____	IWIAAH
46. _____	_____	ANSKAS
47. _____	_____	GORNEO
48. _____	_____	KSAAAL
49. _____	_____	AXSTE
50. _____	_____	IMISRSUO

Abbreviations

SC	AK	MD	ND	FL
IA	ID	KY	RI	TN
NC	MI	MS	GA	MN
OH	IL	WA	AR	SD
MA	CO	CT	WI	PA
WV	MT	HI	LA	OK
NH	MO	KS	VT	UT
VA	NE	OR	WY	ME
NM	NY	AZ	NV	AL
NJ	TX	CA	DE	IN

Supplement 6-1-2 (answer key)

Scrambled States

1. Vermont	VT		26. Indiana	IN
2. South Dakota	SD		27. Arkansas	AR
3. Washington	WA		28. Michigan	MI
4. Iowa	IA		29. Connecticut	CT
5. Minnesota	MN		30. Tennessee	TN
6. Virginia	VA		31. California	CA
7. Colorado	CO		32. New Mexico	NM
8. Georgia	GA		33. Idaho	ID
9. North Carolina	NC		34. South Carolina	SC
10. New Jersey	NJ		35. Florida	FL
11. Ohio	OH		36. Wyoming	WY
12. Maryland	MD		37. New Hampshire	NH
13. West Virginia	WV		38. Rhode Island	RI
14. Wisconsin	WI		39. Louisiana	LA
15. Nebraska	NE		40. New York	NY
16. Mississippi	MS		41. Oklahoma	OK
17. North Dakota	ND		42. Delaware	DE
18. Alabama	AL		43. Montana	MT
19. Illinois	IL		44. Nevada	NV
20. Kentucky	KY		45. Hawaii	HI
21. Utah	UT		46. Kansas	KS
22. Maine	ME		47. Oregon	OR
23. Arizona	AZ		48. Alaska	AK
24. Pennsylvania	PA		49. Texas	TX
25. Massachusetts	MA		50. Missouri	MO

Compass Roses and Other Tools

Fill in the cardinal and intermediate directions on the two compass roses below.

Follow your teacher's directions and turn this square into another direction-finding tool.

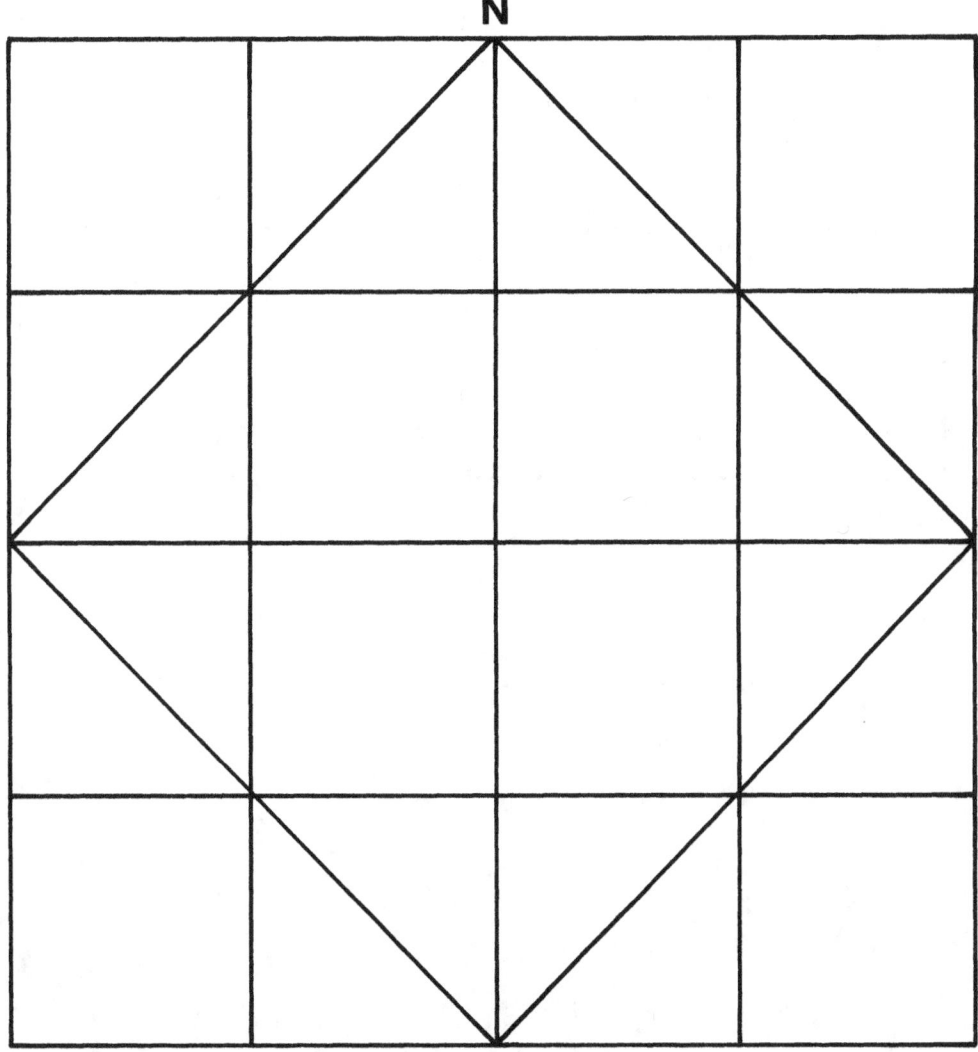

Compass Roses and Other Tools

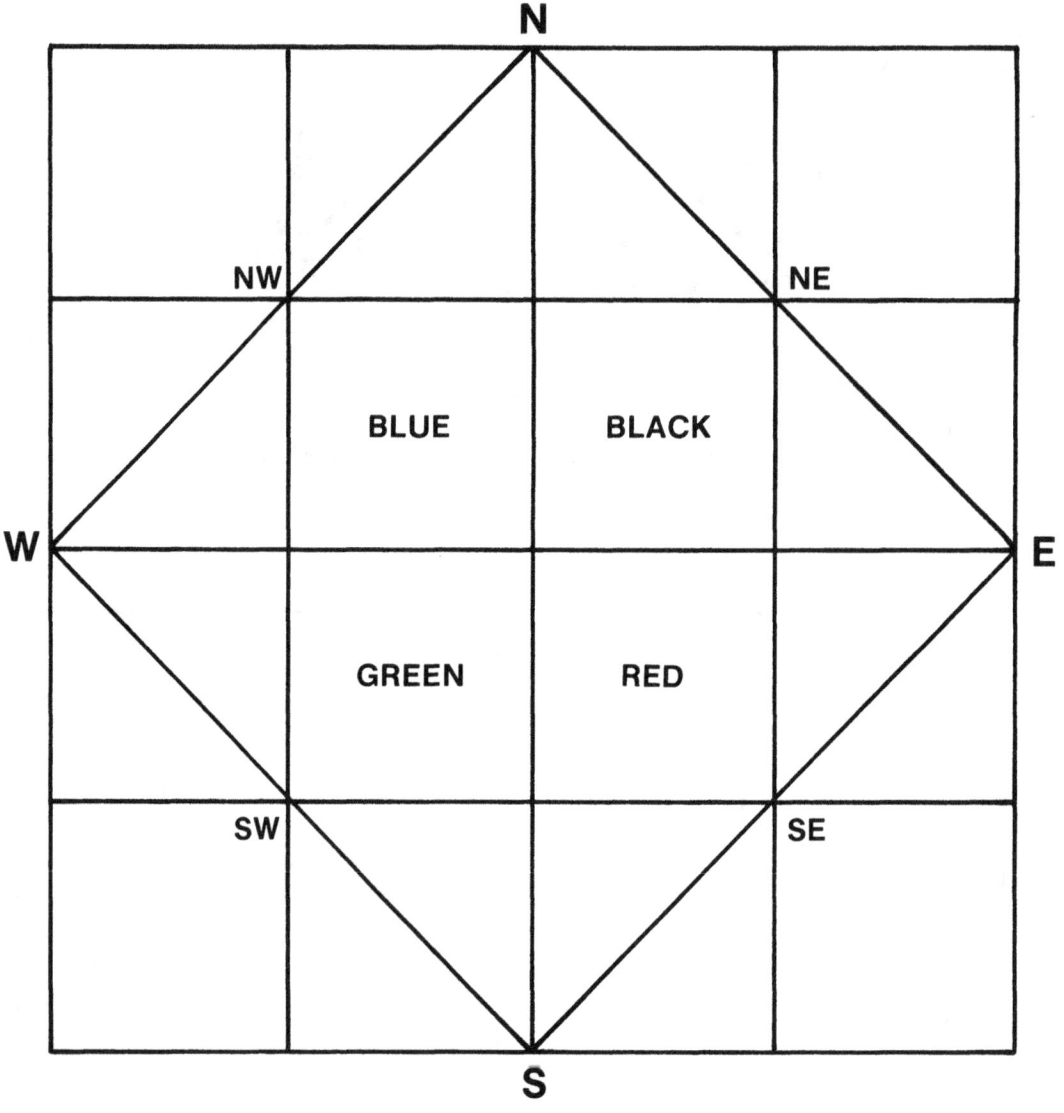

108

Tracing Routes

Use a U.S. map to trace the routes described below. The route in Part 1 will form the outline of an important American symbol. What is it?

Part 1

1. Draw a line from Dallas, Texas, to Minneapolis, Minnesota.
2. Draw a line from Minneapolis, Minnesota, to Birmingham, Alabama.
3. Draw a line from Birmingham, Alabama, to Denver, Colorado.
4. Draw a line from Denver, Colorado, to Columbus, Ohio.
5. Draw a line from Columbus, Ohio, to Dallas, Texas.

What American symbol did you draw? _____

Use your second copy of the U.S. map to complete this part of the activity.

Part 2 Draw a line connecting the following 8 air routes (A thru H). Each route will spell out a letter. Put the letters together to spell out a mystery message.

A.

1. Start where the borders of Canada, Washington, and Idaho meet.
2. Go west to Seattle, Washington.
3. Proceed south to Eureka, California, on the Pacific Ocean.
4. Fly east to where Oregon, Nevada, and Idaho meet.
5. Continue north to Boise, Idaho.
6. Now head west to Bend, Oregon.

What letter have you formed? _____

B.

7. Start at Butte, Montana.
8. Fly south to Salt Lake City, Utah.
9. Head east to Greeley, Colorado.
10. Land at Rapid City, South Dakota..
11. Return to Butte, Montana.

What letter have you formed? _____

Tracing Routes

C.

12. Start your third flight at Sioux Falls, South Dakota.
13. Head south to Wichita, Kansas.
14. Fly northeast to St. Louis, Missouri.
15. Navigate your aircraft to Madison, Wisconsin.
16. Return to Sioux Falls, South Dakota.

What letter have you formed? _____

D.

17. Begin at Detroit, Michigan.
18. Head south to Lexington, Kentucky.
19. Fly east to Richmond, Virginia.
20. Fly northeast to Dover, Delaware.
21. Land at Newark, New Jersey.
22. Fly northwest to Binghamton, New York.
23. End your flight back at Detroit, Michigan.

What letter have you formed? _____

E.

24. Begin this flight at San Francisco, California.
25. Fly southeast to Los Angeles, California.
26. Fly northeast to Las Vegas, Nevada.
27. Head southeast to Phoenix, Arizona.
28. End your flight at Grand Junction, Colorado.

What letter have you formed? _____

F.

29. Start at Albuquerque, New Mexico.
30. Fly south to El Paso, Texas.
31. Navigate east to Waco, Texas.
32. Head north to Oklahoma City, Oklahoma.
33. Return your aircraft to Albuquerque, New Mexico.

What letter have you formed? _____

Tracing Routes

G.

34. Take off from Houston, Texas.
35. Land at Fort Smith, Arkansas.
36. Fly east to Chattanooga, Tennessee.
37. Head southwest to Shreveport, Louisiana.
38. Fly southeast to Mobile, Alabama.

What letter have you formed? _____

H.

39. Start at Greensboro, North Carolina.
40. Fly south to Augusta, Georgia.
41. Head your plane to Tallahassee, Florida.
42. Return to Augusta, Georgia.
43. Fly northeast to Wilmington, North Carolina.
44. Return to Augusta, Georgia for refueling.
45. Fly southeast for a well-deserved vacation in Orlando, Florida.

What letter have you formed? _____

What message did you spell out? _____

Hidden Words

Let's see how good a detective you can be. In the activity below you will be asked to find names of countries hidden within certain words. For example, can you find an Asian country hidden in the word *rain*? If you rearrange the letters, you will find the country of *Iran*.

Look at each word in the list of words below and the clue that you are given. Rearrange the letters in each word so that it becomes the name of a state or country. Can you do them all?

Word	State or Country	Clue
1. pairs		a city in France
2. pure		a country in South America
3. taxes		a state in the United States
4. pains		a country in Europe
5. Nero		a city in Nevada
6. animal		a city in the Philippines
7. rumba		a country in Asia
8. mail		a city in Peru
9. more		a city in Italy
10. solo		a city in Norway
11. also		a country in S.E. Asia
12. serial		a country in the Middle East
13. reign		a country in Africa

Can you create some of your own?

112

Map Search

Find the places described below on a world map. Write their names in the spaces provided.

1. The names of the five Great Lakes. Write them in order so the first letters of their names spell out the word HOMES:

   _____   _____

   _____   _____

   _____

2. An island southeast of a large peninsula in south Asia.

   _____

3. An island west of a larger island northwest of France.

   _____

4. A canal between two continents. _____

5. A sea south of the Ural Mountains and north of Iran.

   _____

6. The world's northernmost bay. _____

7. A lake located at the equator. _____

8. Three seas having colors in their names. _____

   _____   _____

9. Five straits that separate continents. _____

   _____   _____

   _____   _____

10. A peninsula southeast of the Bay of Bengal. _____

Name_____

Map Search

11. Two countries that make up the Iberian Peninsula in southwest Europe.

   _____   _____

12. An island east of the world's second largest continent.

   _____

13. Four deserts whose first letters spell out the word *sand*.

   _____   _____

   _____   _____

14. Twenty-five landlocked countries (countries having no seacoast).

   South America: _____   _____

   Europe: _____   _____

   _____   _____

   Asia: _____   _____

   _____   _____

   Africa: _____   _____

   _____   _____

   _____   _____

   _____   _____

   _____   _____

   _____

15. The world's highest and lowest points (on land). _____

   _____

Supplement 6-2-2 (answer key)

Map Search

1. Huron, Ontario, Michigan, Erie, Superior
2. Sri Lanka
3. Ireland
4. Panama Canal
5. Caspian Sea
6. Baffin Bay
7. Lake Victoria
8. White Sea, Black Sea, Red Sea, Yellow Sea
9. Bering Strait, Strait of Gibraltar, Strait of the Dardanelles, Torres Strait, Strait of Bab-el-Mandeb
10. Malay Peninsula
11. Spain and Portugal
12. Madagascar
13. Sahara or Sonora, Atacama, Nubian, Death Valley
14. South America—Bolivia, Paraguay; Europe—Hungary, Switzerland, Austria, Czechoslovakia; Asia—Nepal, Afghanistan, Bhutan, Mongolia; Africa—Zambia, Botswana, Zimbabwe, Zaire, Chad, Mali, Burkina Faso, Malawi, Rwanda, Burundi, Lesotho, Swaziland, Central African Republic, Uganda, Niger
15. Mt. Everest and Death Valley.

What's in a Name?

In each clue below, fill in the missing letters to find a place whose name includes a boy's name, girl's name, color, a word describing size, or an animal. Use a map or atlas to find each place.

Example: Boy's Name
_ h _ _ _ _ _ to _ , South Carolina is *Charles*ton, South Carolina

Boy's Names

1. N _ _ ma _ , Oklahoma

2. _ arv _ _ , Illinois

3. _ _ ll _ a _ sp _ _ t, Pennsylvania

4. _ t. P _ _ _ , Minnesota

5. A _ _ o _ , Illinois

6. _ _ r _ , Indiana

7. Pi _ _ _ e, South Dakota

8. _ or _ _ rt _ u _ , Texas

9. _ o _ r _ e, Louisiana

10. _ ac _ _ o _ v _ l _e, Florida

11. _ _ u _ s _ i _ _ _ , Kentucky

12. S _ _ _ os _ , California

13. S _ . _ os _ _ _ , Missouri

14. _ va _ _ v _ _ _ e, Indiana

15. P _ _ e _ sb _ _ _ , Virginia

Girl's Names

16. _ _ n _ r _ _ r, Michigan

17. S _ n _ _ _ ar _ _ _ a, California

18. _ e _ en _ , Montana

19. _ l _ z _ _ _ t _ , New Jersey

20. _ l _ r _ _ c _ , South Carolina

21. _ u _ u _ t _ , Maine

22. _ i _ g _ n _ _ , Minnesota

23. B _ _ tr _ _ _ , Nebraska

Colors

24. _ ra _ _ _ , New Jersey

25. _ _ e _ n _ a _ , Wisconsin

26. B _ _ ef _ _ _ _ , West Virginia

27. _ ro _ _ s _ i _ _ e, Texas

28. A _ b _ _ _ , New York

29. _ e _ l _ _ _ s, California

What's in a Name?

30. _ _ ac _ _ _ ll _ , South Dakota

31. _ o _ _ s _ _ _ o, South Carolina

32. _ i _ _ _ t, Louisiana

33. _ r _ _ l _ n _ , Michigan

Size

34. _ ra _ _ F _ _ k _ , North Dakota

35. B _ g S _ _ _ _ g, Texas

36. _ o _ g _ _ a _ h, California

37. _ _ tt _ e _ o _ k, Arkansas

38. G _ _ _ t _ _ lls, Montana

39. _ ho _ _ H _ l _ _ , New Jersey

Animals

40. B _ _ _ e _ r _ i _ y, California

41. E _ _ C _ t _ , Oklahoma

42. _ ar _ _ o _ , Maine

43. _ e _ v _ _ t _ _ , Oregon

44. _ _ xb _ _ o, Massachusetts

45. H _ _ s _ C _ _ _ k, Colorado

46. _ u _ _ a _ o, New York

47. D _ _ r _ i _ _ d _ ea _ h, Florida

48. C _ _ C _ e _ k, Oregon

Supplement 7-2-2 (answer key)

What's in a Name?

1. Norman, Oklahoma
2. Harvey, Illinois
3. Williamsport, Pennsylvania
4. St. Paul, Minnesota
5. Alton, Illinois
6. Gary, Indiana
7. Pierre, South Dakota
8. Port Arthur, Texas
9. Monroe, Louisiana
10. Jacksonville, Florida
11. Louisville, Kentucky
12. San Jose, California
13. St. Joseph, Missouri
14. Evansville, Indiana
15. Petersburg, Virginia
16. Arin Arbor, Michigan
17. Santa Barbara, California
18. Helena, Montana
19. Elizabeth, New Jersey
20. Florence, South Carolina
21. Augusta, Maine
22. Virginia, Minnesota
23. Beatrice, Nebraska
24. Orange, New Jersey
25. Green Bay, Wisconsin
26. Bluefield, West Virginia
27. Brownsville, Texas
28. Auburn, New York
29. Redlands, California
30. Black Hills, South Dakota
31. Goldsboro, South Carolina
32. Violet, Louisiana
33. Grayling, Michigan
34. Grand Forks, North Dakota
35. Big Spring, Texas
36. Long Beach, California
37. Little Rock, Arkansas
38. Great Falls, Montana
39. Short Hills, New Jersey
40. Big Bear City, California
41. Elk City, Oklahoma
42. Caribou, Maine
43. Beaverton, Oregon
44. Foxboro, Massachusetts
45. Horse Creek, Colorado
46. Buffalo, New York
47. Deerfield Beach, Florida
48. Cow Creek, Oregon

Cities Having the Same Name

Each foreign city listed below has at least one sister city located in the United States having the same name.

Using an atlas, find the state (or states) having a city with the same name. Locate both the U.S. and foreign cities on a map.

Foreign City	U.S. City and State(s)
1. Aberdeen, Scotland	
2. Dover, England	
3. Hamburg, Germany	
4. Venice, Italy	
5. Valparaiso, Chile	
6. Geneva, Switzerland	
7. Salem, India	
8. Cairo, Egypt	
9. Canton, China	
10. Trinidad, Cuba	
11. London, England	
12. Vienna, Austria	
13. Monrovia, Liberia	
14. Waterloo, Belgium	
15. San Rafael, Argentina	
16. Birmingham, England	
17. Laredo, Spain	
18. Paris, France	

Cities Having the Same Name

Foreign City	U.S. City and State(s)
19. Belfast, Ireland	
20. Amsterdam, Netherlands	
21. Berlin, Germany	
22. Bethlehem, Jordan	
23. Florence, Italy	
24. Rome, Italy	
25. Melbourne, Australia	
26. Naples, Italy	
27. Bayonne, France	
28. Alexandria, Egypt	
29. Toledo, Spain	
30. Stockholm, Sweden	
31. Athens, Greece	
32. Vancouver, Canada	
33. Moscow, USSR	
34. Odessa, USSR	
35. Lyons, France	
36. Ottawa, Canada	
37. Freeport, Bahamas	
38. Warsaw, Poland	
39. Dublin, Ireland	

Supplement 8-2-2 (answer key)

Cities Having the Same Name

Note: Students will find at least one but not necessarily all of the states listed for each answer. Extent of answer depends on atlas used. You might simplify grading by ensuring that all students use the same atlas.

1. SD, WA, ID, KY, MD, MI, NC
2. FL, GA, ID, IL, IN, KS, KY, MA, MN, MO, NC, ND, NH, NJ, OH
3. AL, AK, CT, IL, IA, MI, MN, MS, NJ, NY, PA
4. OH, UT, CA, FL, IL, LA
5. FL, IN, NE
6. AL, FL, GA, ID, IL, IN, IA, KY, MN, NY, NE, OH, PA, TX
7. AL, AK, CT, FL, IL, IN, IA, KY, ME, MA, MD, MI, MO, NH, NJ, NM, NY, NE, OH, OR, SC, SD, UT, VA, VT, WV, WI
8. GA, IL, FL, KS, MO, NY, NE, OH, OK, WV
9. CT, GA, IL, IN, KS, KY, ME, MA, MN, MS, MO, NC, NJ, NY, OH, OK, PA, SD, TX, WI
10. CA, CO, TX, WA
11. AK, ID, KY, MN, OH, OR, TX, WI
12. GA, IL, ID, LA, ME, MD, MO, NJ, OH, SD, VA, WV
13. AL, CA, IN, MD
14. AL, AK, IL, IN, IA, KS, MT, NY, NE, OH, OR, SC, WI
15. CA, NM
16. AL, IA, MI, MO, NJ, OH, PA
17. MO, MT, TX
18. AK, ID, IL, IA, KY, ME, MI, MS, MO, OH, SC, TN, TX, VA
19. ME, NY, TN
20. GA, MO, MT, NY, OH
21. CT, GA, IL, MA, MD, ND, NH, NJ, NY, OH, OK, PA, WI
22. CT, GA, IN, IA, KY, MD, MS, NH, PA, WV
23. AL, AZ, AK, CO, GA, IL, IN, KS, KY, MA, MN, MS, MO, MT, NC, NJ, NY, OR, PA, SC, SD, TN, TX, VT, WA, WI
24. GA, IL, IN, IA, ME, MS, MO, NY, OH, OR, PA, TN, WI
25. AK, FL, IA, KY, MO, WA
26. FL, ID, IL, ME, NY, SD, TX
27. NJ
28. AL, IN, KY, LA, MN, MO, NH, NE, OH, PA, SD, TN, VA
29. CO, GA, IL, IA, OH, OR, WA
30. ME, NJ, SD, WI
31. AL, AK, GA, IL, ID, LA, ME, MI, NY, OH, PA, TN, TX, WV, WI
32. WA
33. AK, ID, IN, IA, KS, KY, MS, OH, PA, TN, VT
34. NE, TX, WA
35. CO, GA, IL, ID, KS, KY, MI, MO, NJ, NY, NE, OH, OR, SD, WI
36. IL, KS, MN, OH
37. FL, IL, IN, KS, ME, MI, MN, NY, OH, PA, TX
38. IL, IN, KY, MN, MO, NC, ND, NY, OH
39. AK, CA, GA, IN, KY, MD, MI, MS, NC, NH, OH, PA, TX, VA

What Is Scale?

The scale of a map or drawing shows the relationship between the size of the real object and the size of the drawing. For example, imagine an L-shaped block. This drawing shows the block at its actual size. The scale is 1 to 1, or 1:1, or 1. One inch on the drawing stands for one inch on the block.

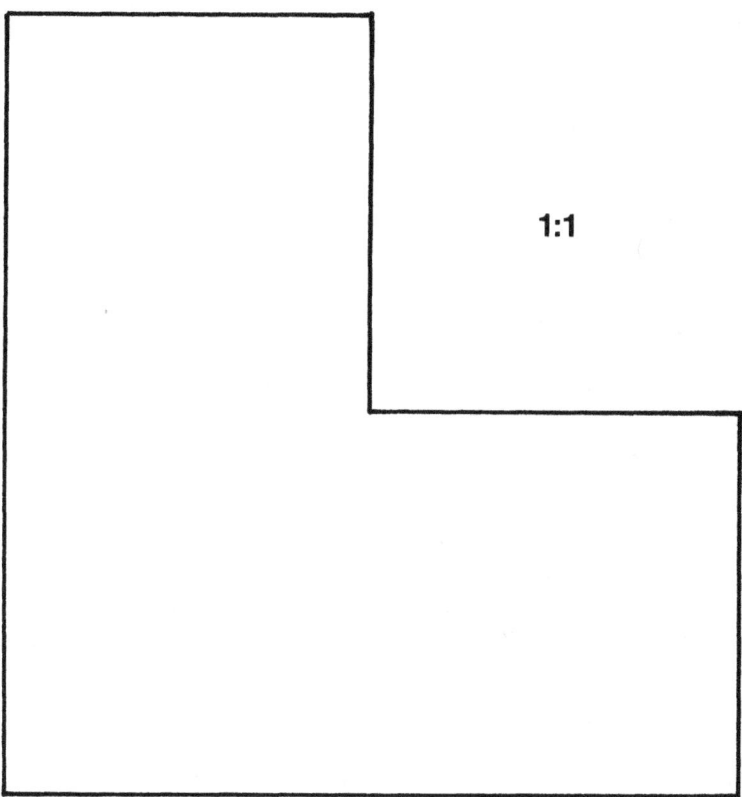

1:1

Now look at the next drawing on page 123. Since 1 inch on this drawing stands for 2 inches on the block, the scale is 1 to 2, or 1:2, or ½.

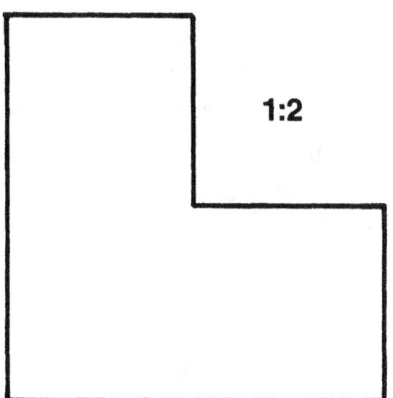

Below is a still smaller drawing. Now 1 inch on the drawing stands for 4 inches on the block so the scale is 1 to 4, or 1:4, or ¼.

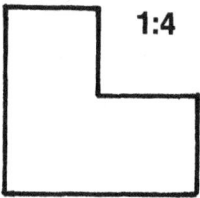

Look at the drawing of the square block. Next to it, make a drawing of the block at a scale of 1 to 2.

Different Kinds of Scales

Map scales can be shown in different ways. One is a ratio, such as 1:3,168,000. This means that 1 inch on the map represents 3,168,000 inches (50 miles) on earth.

Another way of showing scale is through an equation, 1 inch = 50 miles, for example. Here, the equal sign actually means "stands for."

Many maps use a bar scale. A bar scale is marked like a ruler, but the numbers stand for the real distance on earth.

```
L_____J
0          50 miles
```

All the scale examples above show the same scale.

Change each of the scales below to another form.

1.
```
L_____J
0          100 miles
```
Show this scale as an equation.

_____

2.   ½ inch = 25 miles   Show this scale as a bar scale.

3.
```
L_____J
0          50 feet
```
Show this scale as a ratio. _____

Using Scale

Use the map below to answer questions 1-5.

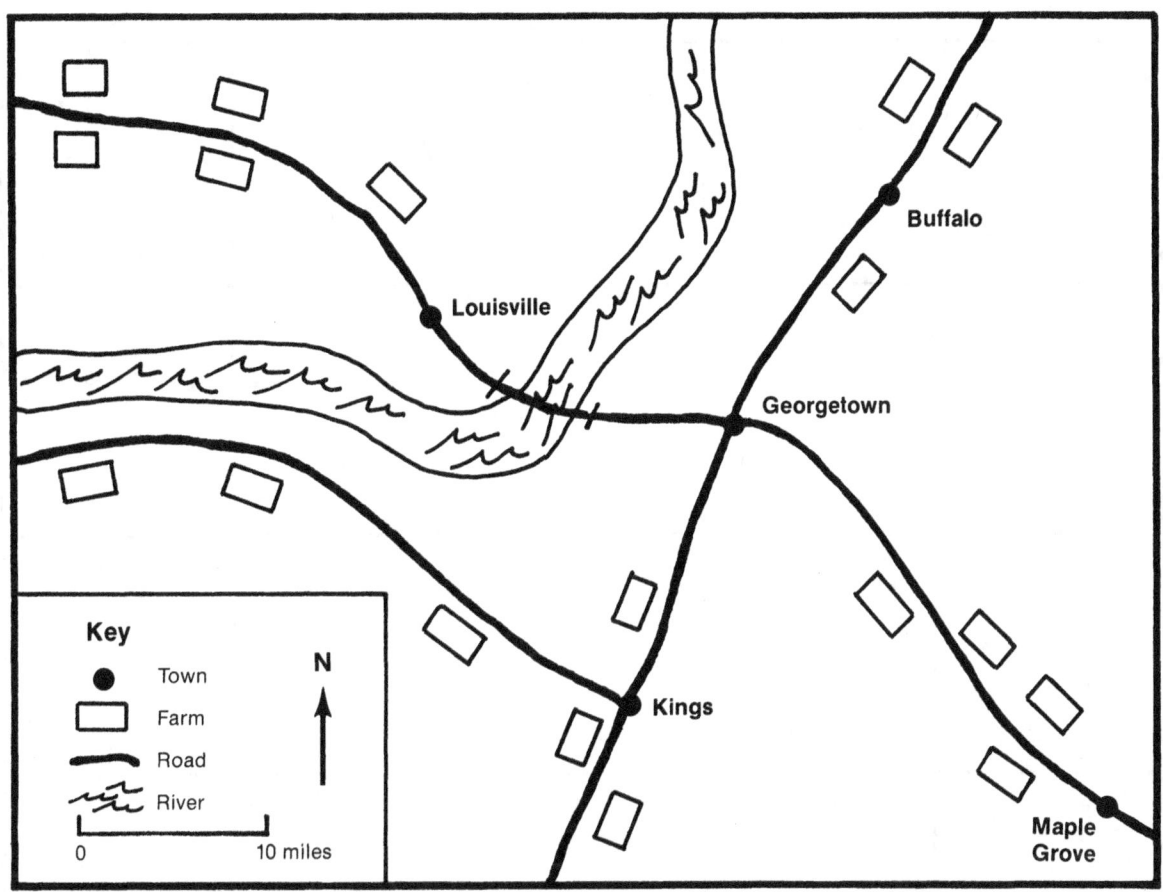

1. How far is it from Buffalo to Georgetown? _____

2. How far is the first farm northwest of Louisville from Georgetown?

   _____

3. Which two towns are farthest apart? _____

4. If you lived in Buffalo, would it take longer to get to Maple Grove by driving

   through Georgetown or by going cross-country? _____

   About how much longer? _____

5. Imagine that you live in Georgetown. You must visit all four of the other
   towns on business. You must travel by road. How far will you travel?

   _____

    125

A Grid

	A	B	C	D	E	
1						1
2						2
3						3
4						4
5						5
	A	B	C	D	E	

126

Supplement 1-4-2 (answer key)

A Grid

	A	B	C	D	E
1	First name				Blue
2		Favorite color		Blue	
3	Red		Blue		
4	Red	Blue		Favorite color	
5	Red	Red	Red		Last name

Location Coordinates and the World Map

Use a world map to find each set of latitude/longitude coordinates in the list below. In column C write the name of the country in which the grid point is found. In column D write the name of the capital city.

A	B	C	D
Latitude	Longitude	Country	Capital City
1. 40°N	90°W		
2. 60°N	60°E		
3. 30°S	150°E		
4. 20°S	50°W		
5. 1°N	37°E		
6. 39°N	32°E		
7. 27°N	27°E		
8. 53°N	17°E		
9. 44°N	11°E		
10. 23°N	77°E		
11. 40°N	5°W		
12. 37°N	110°E		
13. 46°N	100°E		
14. 36°S	65°W		
15. 36°N	134°E		
16. 39°N	22°E		
17. 28°S	25°E		
18. 22°N	79°W		

Location Coordinates and the World Map

A	B	C	D
Latitude	**Longitude**	**Country**	**Capital City**
19. 29°N	1°E		
20. 56°N	4°W		
21. 5°N	75°W		
22. 62°N	6°E		
23. 20°N	100°W		
24. 35°N	51°E		

Supplement 4-4-2 (answer key)

Location Coordinates and the World Map

1. United States, Washington, D.C.
2. Soviet Union, Moscow
3. Australia, Canberra
4. Brazil, Brasilia
5. Kenya, Nairobi
6. Turkey, Ankara
7. Egypt, Cairo
8. Poland, Warsaw
9. Italy, Rome
10. India, New Delhi
11. Spain, Madrid
12. China, Beijing
13. Mongolia, Ulaanbaatar
14. Argentina, Buenos Aires
15. Japan, Tokyo
16. Greece, Athens
17. South Africa, Cape Town
18. Cuba, Havana
19. Algeria, Algiers
20. Scotland, Edinburgh
21. Colombia, Bogota
22. Norway, Oslo
23. Mexico, Mexico City
24. Iran, Teheran

Name_____

Coordinates and Capitals

Each set of latitude/longitude coordinates below describes a grid point near a state capital. Use a map to help you locate each state capital. Write the name of the capital in column C and the state abbreviation in column D. Then use reference materials to help you select the correct state nickname from the list on page 134.

A and B Latitude/Longitude	C Capital	D State	E Nickname
1. 32°N - 86°W			
2. 58°N - 134°W			
3. 33°N - 112°W			
4. 34°N - 92°W			
5. 21°N - 158°W			
6. 47°N - 123°W			
7. 45°N - 123°W			
8. 40°N - 122°W			
9. 43°N - 116°W			
10. 39°N - 119°W			
11. 40°N - 112°W			
12. 46°N - 112°W			
13. 39°N - 105°W			
14. 41°N - 104°W			
15. 35°N - 106°W			
16. 30°N - 97°W			

Name_____

Coordinates and Capitals

A and B Latitude/Longitude	C Capital	D State	E Nickname
17. 35°N - 97°W			
18. 39°N - 95°W			
19. 40°N - 96°W			
20. 44°N - 100°W			
21. 47°N - 100°W			
22. 45°N - 93°W			
23. 41°N - 93°W			
24. 38°N - 92°W			
25. 30°N - 91°W			
26. 32°N - 90°W			
27. 36°N - 86°W			
28. 38°N - 85°W			
29. 39°N - 89°W			
30. 43°N - 89°W			
31. 42°N - 84°W			
32. 39°N - 86°W			
33. 40°N - 83°W			
34. 33°N - 84°W			
35. 30°N - 84°W			
36. 34°N - 81°W			

Coordinates and Capitals

A and B Latitude/Longitude	C Capital	D State	E Nickname
37. 35°N - 78°W			
38. 37°N - 77°W			
39. 38°N - 81°W			
40. 39°N - 76°W			
41. 39°N - 75°W			
42. 40°N - 74°W			
43. 40°N - 77°W			
44. 43°N - 74°W			
45. 41°N - 72°W			
46. 42°N - 71°W			
47. 44°N - 72°W			
48. 43°N - 71°W			
49. 44°N - 69°W			
50. 42°N - 71°W			

Coordinates and Capitals
State Nicknames

Garden State	Little Rhody - Ocean State
Aloha State	Centennial State
Treasure State	Cornhusker State
Show Me State	Heart of Dixie - Cotton State
Empire State	Old Line State
Badger State	Pine Tree State
Magnolia State	Bluegrass State
Golden State	Land of Enchantment
Lone Star State	First State
Hawkeye State	Constitution State - Nutmeg State
Pelican State	Empire State of the South - Peach State
Gem State	Green Mountain State
Sunshine State	Great Lake State - Wolverine State
Buckeye State	Tar Heel State
Volunteer State	Bay State
Old Dominion	The Last Frontier
Keystone State	Land of Opportunity
Hoosier State	Grand Canyon State
Mountain State	Evergreen State
Granite State	Sagebrush State - Battle Born State
Palmetto State	Sunflower State
Equality State	North Star State - Gopher State
Sooner State	Coyote State - Sunshine State
Beaver State	Sioux State - Flickertail State
Beehive State	Prairie State

Supplement 5-4-2 (answer key)

Coordinates and Capitals

1. Montgomery, AL, Heart of Dixie/ Cotton State
2. Juneau, AK, The Last Frontier
3. Phoenix, AZ, Grand Canyon State
4. Little Rock, AR, Land of Opportunity
5. Honolulu, HI, Aloha State
6. Olympia, WA, Evergreen State
7. Salem, OR, Beaver State
8. Sacramento, CA, Golden State
9. Boise, ID, Gem State
10. Carson City, NV, Sagebrush State/ Battle Born State
11. Salt Lake, City, UT, Beehive State
12. Helena, MT, Treasure State
13. Denver, CO, Centennial State
14. Cheyenne, WY, Equality State
15. Santa Fe, NM, Land of Enchantment
16. Austin, TX, Lone Star State
17. Oklahoma City, OK, Sooner State
18. Topeka, KS, Sunflower State
19. Lincoln, NE, Cornhusker State
20. Pierre, SD, Coyote State/ Sunshine State
21. Bismarck, ND, Sioux State/ Flickertail State
22. St. Paul, MN, North Star/ Gopher State
23. Des Moines, IA, Hawkeye State
24. Jefferson City, MO, Show Me State
25. Baton Rouge, LA, Pelican State
26. Jackson, MS, Magnolia State
27. Nashville, TN, Volunteer State
28. Frankfort, KY, Bluegrass State
29. Springfield, IL, Prairie State
30. Madison, WI, Badger State
31. Lansing, MI, Great Lake State/ Wolverine State
32. Indianapolis, IN, Hoosier State
33. Columbus, OH, Buckeye State
34. Atlanta, GA, Empire State of the South/Peach State
35. Tallahassee, FL, Sunshine State
36. Columbia, SC, Palmetto State
37. Raleigh, NC, Tar Heel State
38. Richmond, VA, Old Dominion
39. Charleston, WV, Mountain State
40. Annapolis, MD, Old Line State
41. Dover, DE, First State
42. Trenton, NJ, Garden State
43. Harrisburg, PA, Keystone State
44. Albany, NY, Empire State
45. Hartford, CT, Constitution State/ Nutmeg State
46. Providence, RI, Little Rhody/ Ocean State
47. Montpelier, VT, Green Mountain State
48. Concord, NH, Granite State
49. Augusta, ME, Pine Tree State
50. Boston, MA, Bay State.

What's on the Menu?

Each of the places located at the latitude/longitude coordinates listed below has a name associated with a food or beverage. Use a map or atlas to help you locate each place. Write the name in the column provided.

Latitude	Longitude	Name Associated with Food or Beverage
1. 48.10° N	89.07° W	
2. 42.35° N	88.53° W	
3. 45.30° N	115.45° W	
4. 51.16° N	92.46° W	
5. 35.55° N	87.40° W	
6. 30.07° N	93.44° W	
7. 44.30° N	11.18° E	
8. 28.20° N	80.35° W	
9. 48.25° N	108.45° W	
10. 47.21° N	119.09° W	
11. 44.10° N	74.50° W	
12. 39.30° N	74.35° W	
13. 48.15° N	98.35° W	
14. 53.06° N	2.01° W	
15. 31.23° N	34.30° E	
16. 53.34° N	10.02° E	
17. 38.12° N	103.42° W	
18. 33.40° N	112.01° W	
19. 34.00° N	117.51° W	

Name_____

What's on the Menu?

Latitude	Longitude	Name Associated with Food or Beverage
20. 44.14° N	88.27° W	
21. 35.32° N	81.22° W	
22. 42.10° N	77.05° W	
23. 49.26° N	94.14° W	
24. 32.56° N	97.05° W	
25. 38.15° N	83.10° W	

Supplement 6-4-2 (answer key)

What's on the Menu

1. Pie Island, Canada
2. Sandwich, Illinois
3. Salmon River Mountains, Idaho
4. Trout Lake, Canada
5. Duck River, Tennessee
6. Orange, Texas
7. Bologna, Italy
8. Cocoa Beach, Florida
9. Milk River, Montana
10. Crab Creek, Washington
11. Cranberry Lake, New York
12. Egg Harbor, New Jersey
13. Sweetwater Lake, North Dakota
14. Leek, England
15. Beer River, Israel
16. Hamburg, Germany
17. Sugar City, Colorado
18. Salt River Indian Reservation, Arizona
19. Walnut, California
20. Appleton, Wisconsin
21. Cherryville, North Carolina
22. Corning, New York
23. Whitefish Bay, Canada
24. Grapevine, Texas
25. Olive Hill, Kentucky

Making a Flight Plan

Pretend you are a business executive about to leave on a long trip with many stopovers. You are laying your flight plan out on a U.S. map before you leave. Begin at El Paso, Texas, where you are now. Then follow the directions below, drawing lines on the map connecting each stopover of your journey. When you have connected all the stopovers you have made and the trip is completed, you will have drawn the outline of a mode of transportation. What is it?

1. Start your trip at your home base in El Paso, Texas. Fly to Phoenix, Arizona.
2. From Phoenix, fly due east to a secret airport at 33°N, 103°W.
3. Fly north to the 37°N parallel. Fly east along the parallel until you get to the Kansas-Missouri border.
4. Fly north to Kansas City, Missouri.
5. Proceed to St. Louis, Missouri.
6. Fly south until you get to the 37°N parallel again.
7. Fly east along the parallel until you arrive at a point about 30 miles north of Winston-Salem, North Carolina.
8. Fly south to Columbia, South Carolina.
9. Make a quick trip to Wilmington, North Carolina.
10. Fly to 30°N, 82°W.
11. Continue west to 30°N, 105°W for a secret meeting.
12. Return to your home base in El Paso, Texas.

What form of transportation did you draw by connecting the cities enroute

to and from El Paso, Texas? _____

Name_____

Tallest Buildings in the United States

Below is a list of the cities in the United States that have buildings over 800 feet high. Use an almanac or the encyclopedia to find the names of these buildings. Give their height in feet and the number of stories.

City	Building	Height	Stories
1. Atlanta, GA			
2. Chicago, IL			
3. Dallas, TX			
4. Hartford, CT			
5. Houston, TX			
6. Los Angeles, CA			

Name_____

Tallest Buildings in the United States

City	Building	Height	Stories
7. New York, NY			
8. Philadelphia, PA			
9. Pittsburgh, PA			
10. San Francisco, CA			
11. Seattle, WA			

Supplement 1-5-2 (answer key)

Tallest Buildings in the United States

Note: Accept slight variations in answers, as sources sometimes differ on height.

	City	Building	Height	Stories
1.	Atlanta, GA	IBM Tower	825	50
2.	Chicago, IL	Sears Tower	1,454	110
		Amoco	1,136	80
		John Hancock	1,127	100
		Two Franklin Place	970	65
		311 S. Wacker	959	65
		Two Prudential Plaza	901	64
		AT&T Corporate Center	891	60
		900 N. Michigan Ave.	871	66
		Water Tower Place	859	74
		First National Bank	852	60
3.	Dallas, TX	First Republic Bank Plaza	939	73
4.	Hartford, CT	Cutter Financial Center	878	59
		Oakleaf Building	801	46
5.	Houston, TX	Texas Commerce Tower	1,002	75
		Allied Bank Plaza	992	71
		Transco Tower	901	64
6.	Los Angeles, CA	Library Tower	1,017	73
		First Interstate Bank	858	62
7.	New York, NY	World Trade Center	1,350	110
		Empire State Bldg.	1,414	102
		Chrysler Bldg.	1,046	77
		40 Wall Tower	927	71
		Citicorp Center	914	46
		G.E. Bldg.		
		Rockefeller Center	850	70
		American International	826	67
		Chase Manhattan Plaza	813	60
		Cityspire	813	72
		Pan Am Bldg.	808	59
8.	Philadelphia, PA	One Liberty Place	940	60
		Mellon Bank Center	886	56
		Two Liberty Place	825	60
9.	Pittsburgh, PA	USX Towers	841	64
10.	San Francisco, CA	Transamerica Pyramid	853	48
11.	Seattle, WA	Columbia Seafirst Center	954	76

True or False?

What kind of map would you use to find the following information? Use that map to help you answer *True* or *False* to the following statements.

1. You must travel eastward to get from the Atlantic Ocean to the Pacific Ocean via the Panama Canal. _____

2. New York City and Madrid are located at the same latitude. _____

3. About two-thirds of the world is water. _____

4. The southernmost U.S. city is in Hawaii. _____

5. Reno, Nevada, is farther west than Los Angeles, California. _____

6. The United States and the Soviet Union are only 39 miles apart at the Bering Strait. _____

7. India's coastal waters include a sea, ocean, gulf, bay, and strait. _____

8. Parts of the Soviet Union extend to latitudes as far south as Africa and farther north than Alaska; others have longitudinal positions farther east than New Zealand and about as far west as Greece. _____

9. The United States is about the same distance from east to west as Chile is from north to south. _____

Symbols for Weather Elements and Cloud Types

WEATHER ELEMENTS

∞	HAZE	✳	SNOW
=	MIST	▽	RAIN SHOWERS
≡	FOG	✳▽	SNOW SHOWERS OR FLURRIES
	SMOKE	△▽	HAIL
,	DRIZZLE	⟨	THUNDERSTORM
●	RAIN	S	DUST OR SANDSTORM
●∿	FREEZING RAIN	↓	DRIFTING SNOW

CLOUD TYPES

→	CIRRUS	∿	STRATOCUMULUS
2	CIRROSTRATUS	∠	NIMBOSTRATUS
℘	CIRROCUMULUS	△	CUMULUS
∿	ALTOCUMULUS	⌂	CUMULUS CONGESTUS
∠	ALTOSTRATUS	⊠	CUMULONIMBUS
—	STRATUS	- - -	SCUD

Symbols for Barometric Tendency

RISING, THEN FALLING

RISING, THEN STEADY OR RISING; THEN RISING MORE SLOWLY.

RISING UNSTEADILY, OR UNSTEADY.

RISING STEADILY, OR STEADILY (NOT PLOTTED)

FALLING OR STEADY, THEN RISING; OR RISING, THEN RISING MORE QUICKLY.

} BAROMETER NOW HIGHER THAN, OR SAME AS 3 HOURS AGO.

FALLING, THEN RISING

FALLING, THEN STEADY OR FALLING; THEN FALLING MORE SLOWLY.

FALLING UNSTEADILY, OR UNSTEADY

FALLING STEADILY (NOT PLOTTED)

STEADY OR RISING, THEN FALLING; OR FALLING, THEN FALLING MORE QUICKLY.

} BAROMETER NOW LOWER THAN 3 HOURS AGO.

Beaufort Scale of Winds

SYMBOL	NO.	NAME OF WIND	EFFECTS	MILES PER HR.
◯	0	CALM	SMOKE RISES STRAIGHT UP.	LESS THAN 1
◯—	1	LIGHT AIR	SMOKE DRIFTS GENTLY. WIND VANE STILL.	1-3
◯—	2	LIGHT BREEZE	FACE FEELS WIND. LEAVES RUSTLE.	4-7
◯—	3	GENTLE BREEZE	LEAVES AND SMALL TWIGS MOVE.	8-12
◯—	4	MODERATE BREEZE	STIRS DUST, PAPER AND SMALL BRANCHES.	13-18
◯—	5	FRESH BREEZE	SMALL TREES SWAY. WAVELETS ON LAKES.	19-24
◯—	6	STRONG BREEZE	BRANCHES MOVE. WIND SOUND IN WIRES.	25-31
◯—	7	MODERATE GALE	WHOLE TREES MOVE. WALKING DIFFICULT.	32-38
◯—	8	FRESH GALE	TWIGS BROKEN OFF.	39-46
◯—	9	STRONG GALE	LOOSE CHIMNEYS AND SHINGLES GO.	47-54
◯—	10	WHOLE GALE	TREES UPROOTED.	55-63
◯—	11	STORM	WIDESPREAD DAMAGE.	64-72
◯—	12	HURRICANE	ANYTHING MAY GO.	ABOVE 72

Weather Report from One Station

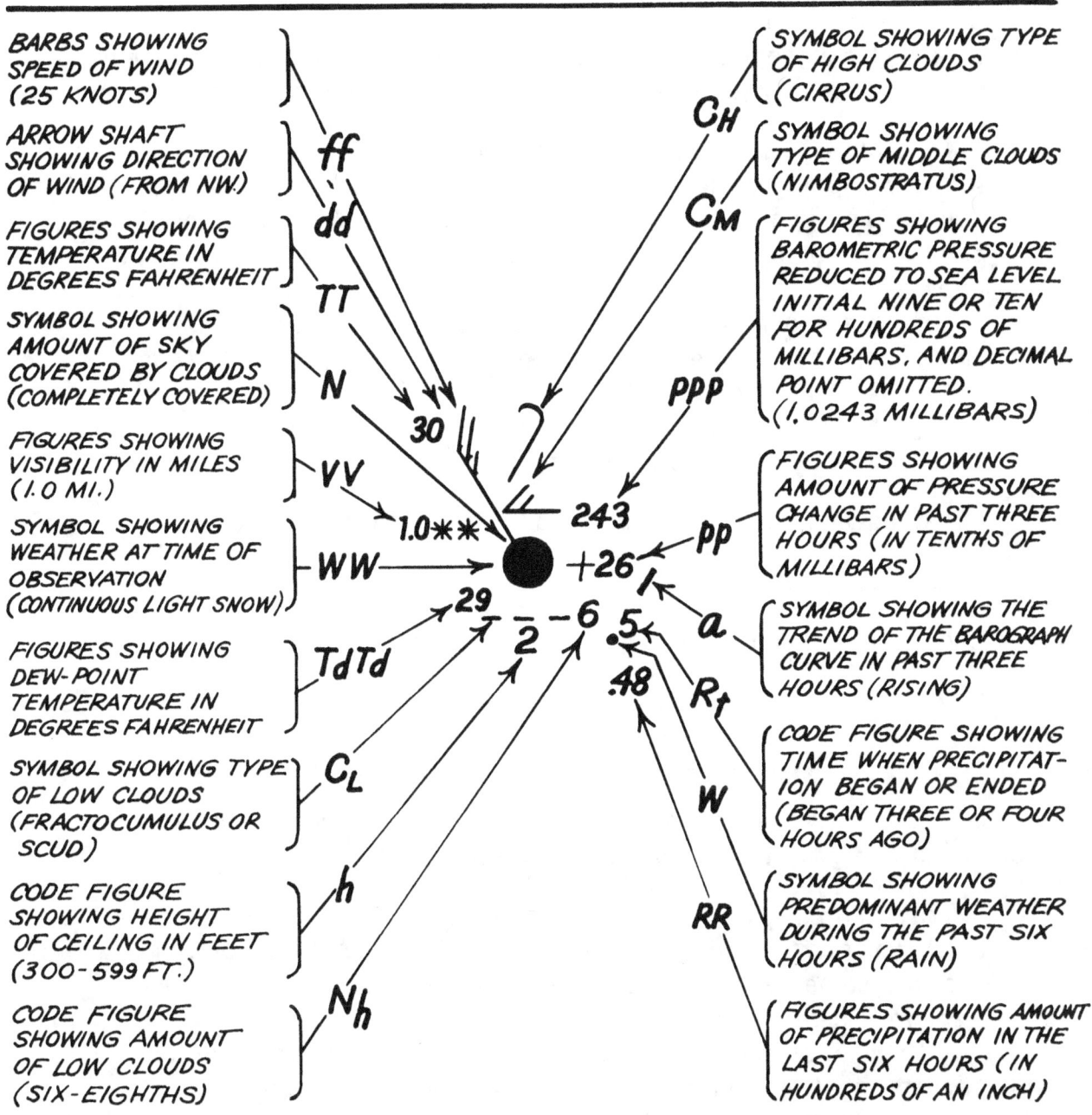

BARBS SHOWING
SPEED OF WIND
(25 KNOTS)

ARROW SHAFT
SHOWING DIRECTION
OF WIND (FROM NW.)

FIGURES SHOWING
TEMPERATURE IN
DEGREES FAHRENHEIT

SYMBOL SHOWING
AMOUNT OF SKY
COVERED BY CLOUDS
(COMPLETELY COVERED)

FIGURES SHOWING
VISIBILITY IN MILES
(1.0 MI.)

SYMBOL SHOWING
WEATHER AT TIME OF
OBSERVATION
(CONTINUOUS LIGHT SNOW)

FIGURES SHOWING
DEW-POINT
TEMPERATURE IN
DEGREES FAHRENHEIT

SYMBOL SHOWING TYPE
OF LOW CLOUDS
(FRACTOCUMULUS OR
SCUD)

CODE FIGURE
SHOWING HEIGHT
OF CEILING IN FEET
(300-599 FT.)

CODE FIGURE
SHOWING AMOUNT
OF LOW CLOUDS
(SIX-EIGHTHS)

ff
dd
TT
N
VV
WW
T_dT_d
C_L
h
N_h

SYMBOL SHOWING TYPE
OF HIGH CLOUDS
(CIRRUS)

SYMBOL SHOWING
TYPE OF MIDDLE CLOUDS
(NIMBOSTRATUS)

FIGURES SHOWING
BAROMETRIC PRESSURE
REDUCED TO SEA LEVEL
INITIAL NINE OR TEN
FOR HUNDREDS OF
MILLIBARS, AND DECIMAL
POINT OMITTED.
(1,0243 MILLIBARS)

FIGURES SHOWING
AMOUNT OF PRESSURE
CHANGE IN PAST THREE
HOURS (IN TENTHS OF
MILLIBARS)

SYMBOL SHOWING THE
TREND OF THE BAROGRAPH
CURVE IN PAST THREE
HOURS (RISING)

CODE FIGURE SHOWING
TIME WHEN PRECIPITAT-
ION BEGAN OR ENDED
(BEGAN THREE OR FOUR
HOURS AGO)

SYMBOL SHOWING
PREDOMINANT WEATHER
DURING THE PAST SIX
HOURS (RAIN)

FIGURES SHOWING AMOUNT
OF PRECIPITATION IN THE
LAST SIX HOURS (IN
HUNDREDS OF AN INCH)

C_H
C_M
PPP
pp
a
R_t
W
RR

A DETAILED WEATHER REPORT FOR ONE WEATHER STATION.
THE DATA PROVIDED BY MANY SUCH STATIONS AT A PRESCRIBED
TIME ARE COMBINED TO MAKE THE SYNOPTIC WEATHER MAP, AN
IMPORTANT BASIS FOR WEATHER FORECASTING.

Making Inferences

As you read the following story, circle the key words that are objects you probably would find in a specific geographic region of the United States.

After studying a geographic regions map of the United States, review the word clues that you circled in the story and decide to which region they are referring.

Anaki could hear the rain hitting the bark walls of her family's wigwam. Slipping on her deerskin moccasins, she walked across the pine needle carpet to look outside. She hoped it would stop raining before it was time to work in the fields that her family had cleared in the forest.

Today they would be picking beans and corn. Potatoes and squash also grew in the fields. Although Anaki liked working in the fields, she did not like getting wet. Just then, she saw her older brother walking through the village toward their wigwam. He carried his spear and several fish. Suddenly Anaki felt hungry.

1. In what region of the United States did the above story occur?

   _____

2. Do you think the story took place in the 1700s or the 1900s? _____

   Why? _____

   _____

Read the following modern story. Once again, circle the words that will help you identify the geographic location of the story.

Lisa's alarm clock rang at 5:00 A.M. Even though the sun had not risen, the desert was already warm. Lisa knew she would be hot when she finished delivering papers on her paper route. She was glad there were no hills on her route.

Pulling on her jeans and a T-shirt, Lisa ran downstairs for a quick bowl of cereal and a glass of juice. The air-conditioned house was quiet. Her mother would not be up for another hour. She did not have to leave for her job at the solar energy company until 7:00 A.M.

Lisa ran down the front steps of her house and jumped on her bicycle. She was off!

1. In what state might this story occur? _____ Why?

   _____

　　　　148

Making Simple Floor Plans

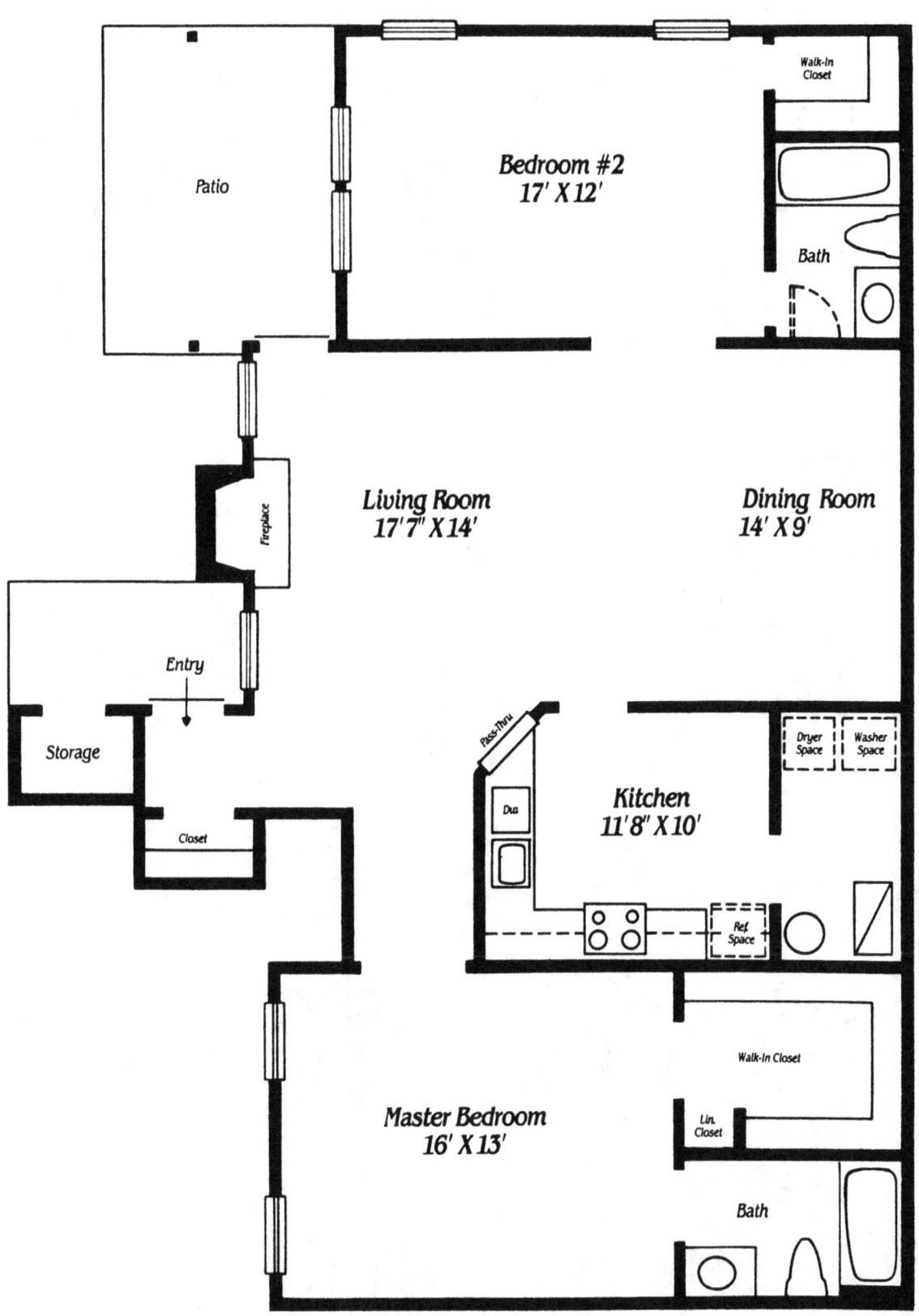

Safety Symbols

What do these symbols mean? Can you find them in your neighborhood?

1. _____   2. _____   3. _____

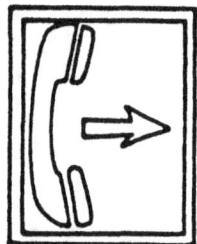

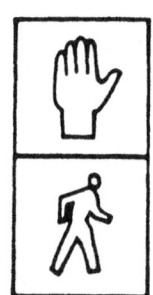

4. _____   5. _____   6. _____

7. _____   8. _____   9. _____   10. _____

150

Supplement 4-1-3 (answer key)

Safety Symbols

1. reserved for the disabled
2. railroad crossing
3. traffic stop
4. public telephone available
5. pedestrian crossing - stop and go
6. pedestrian crosswalk
7. bicycle crossing
8. no bicycles allowed
9. oncoming traffic has right of way
10. no smoking allowed

Stop, Look, and Listen When You Travel

Have you ever found that when you visit a new city or town your "memory picture" of the place fades soon after you return home? This often happens because there is so much to see and do in new places.

To help you learn and remember more about places you visit, take this checklist along the next time you visit another town or city. See how many questions you can answer.

Stop, look, and listen as you explore the city and talk to people. Think of yourself as a reporter trying to discover valuable clues to a story.

1. How did the place get its name? What does it mean?

2. How old is the town or city? Why did it begin where it did?

3. For what is the town or city famous?

4. Are some streets or sections of the city named for geographic features? What are the names?

5. Do some places have names that tell about the history of the community? What are the names?

6. Who were the early settlers? Why did they come there?

7. How many people live there?

8. How is the community changing? What is old? What is new?

9. How do most of the people earn a living?

10. What type of weather do they have throughout the year?

11. What do people do for fun? Are there parks and playgrounds?

12. How do goods and services go in and out of the community?

13. What major roads/highways lead in and out of the city?

14. What are some of the important landmarks?

15. How would you describe the natural environment?

16. What did you like about the community? What did you dislike?

17. How is the town or city like your community? How is it different?

Place Names

Place names tell us about the history and geography of a state. Look at the list of natural features below. For each feature, the name of a state is given. Use a state map to help you find a place name within the state that has, as part of its name, the natural feature listed.

Natural Feature	**Place Name**
1. basin	_____, Utah
2. bay	_____, Wisconsin
3. bayou	_____, Louisiana
4. beach	_____, Florida
5. bluff	_____, Arkansas
6. branch	_____, New Jersey
7. butte	_____, Montana
8. canyon	_____, Arizona
9. cape	_____, North Carolina
10. cavern	_____, New Mexico
11. cliff	_____, New Jersey
12. Continental Divide	_____, New Mexico
13. cove	_____, New York
14. crater	_____, Oregon
15. creek	_____, Michigan
16. dale	_____, Pennsylvania
17. delta	_____, Texas
18. falls	_____, Montana
19. ford	_____, Colorado
20. fork	_____, North Dakota

Name_____

Place Names

Natural Feature	Place Name
21. gap	_____ , New Jersey
22. glacier	_____ , Alaska
23. gulf	_____ , Mississippi
24. harbor	_____ , Maine
25. hill	_____ , South Dakota
26. inlet	_____ , Washington
27. island	_____ , California
28. key	_____ , Florida
29. lake	_____ , Utah
30. marsh	_____ , Wisconsin
31. mesa	_____ , Arizona
32. mountain	_____ , Wyoming
33. ocean	_____ , Maryland
34. pass	_____ , California
35. peak	_____ , Colorado
36. peninsula	_____ , Alaska
37. piedmont	_____ , West Virginia
38. plain	_____ , Texas
39. plateau	_____ , Texas
40. point	_____ , Washington
41. pond	_____ , Maine
42. rapids	_____ , Iowa
43. ridge	_____ , South Dakota

Name_____

Place Names

Natural Feature	Place Name
44. river	_____, Michigan
45. sand (bar)	_____, Oregon
46. sea	_____, California
47. shoal	_____, Alabama
48. shore	_____, Michigan
49. sound	_____, Washington
50. spring	_____, Colorado
51. strait	_____, Washington
52. stream	_____, New York
53. swamp	_____, Georgia
54. valley	_____, North Dakota

Supplement 1-3-3 (answer key)

Place Names

1. Great Basin
2. Green Bay
3. Bayou Cave
4. Palm Beach
5. Pine Bluff
6. Long Branch
7. Butte
8. Grand Canyon
9. Cape Hatteras
10. Carlsbad Cavern
11. Cliffside Park
12. Continental Divide
13. Glen Cove
14. Crater Lake National Park
15. Battle Creek
16. Dale
17. Delta
18. Great Falls
19. Rocky Ford
20. Grand Forks
21. Delaware Water Gap
22. Glacier Bay National Park
23. Gulfport
24. Bar Harbor
25. Black Hills
26. Admiralty Inlet
27. Santa Catalina Island
28. Key West
29. Great Salt Lake
30. Marshfield
31. Mesa
32. Big Horn Mountains
33. Ocean City
34. Tehachapi Pass
35. Pikes Peak
36. Alaska Peninsula
37. Piedmont
38. Plainview
39. Edwards Plateau
40. Point Roberts
41. Great Pond
42. Cedar Rapids
43. Pine Ridge
44. River Rouge
45. Sand
46. Salton Sea
47. Muscle Shoals
48. St. Clair Shores
49. Puget Sound
50. Steamboat Springs
51. Juan de Fuca Strait
52. Valley Stream
53. Okefenokee Swamp
54. Valley City

Natural Geographic Features

archipelago	dale	pass
arm	delta	peak
arroyo	divide	peninsula
atoll	estuary	piedmont
badlands	fall line	plain
bank	fjord	plateau
basin	flood plain	point
bay	foothill	pond
bayou	ford	range
beach	fork	rapids
bluff	gap	ravine
bog	glacier	reef
branch	gorge	ridge
brook	gulch	river
butte	gulf	sandbar
canyon	gully	sand dunes
cape	harbor	sea
cave	hill	shoal
cavern	inlet	shore
channel	island	sound
cliff	isthmus	spit
coast	key	strait
coastal plain	knob	stream
continent	knoll	swamp
continental shelf	lagoon	tableland
coral reef	lake	timberline
cove	ledge	valley
crag	marsh	volcano
crater	mesa	waterfall
creek	mountain	
current	ocean	

Natural Geographic Features

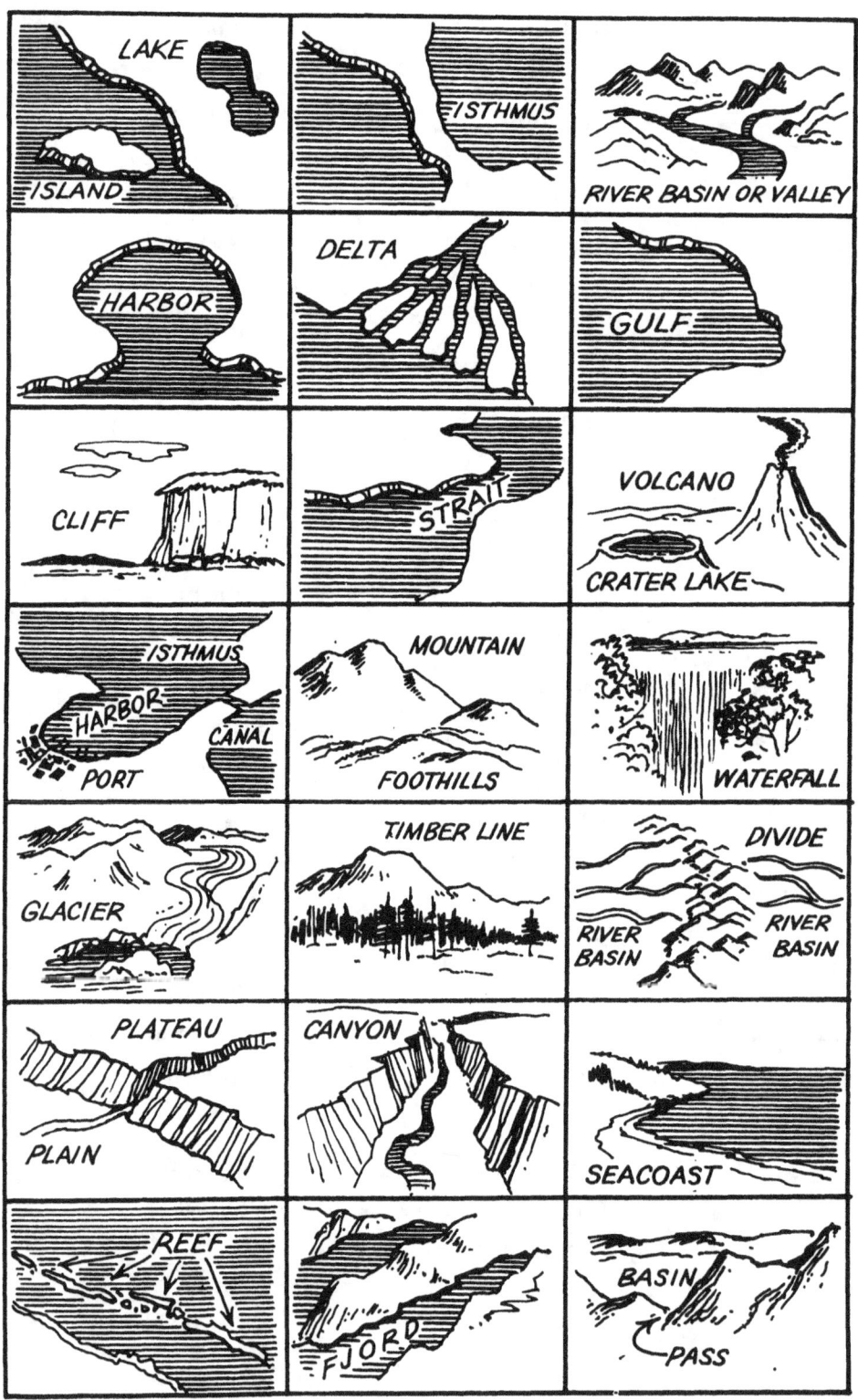

Supplement 1-4-3(c)

Natural Geographic Features

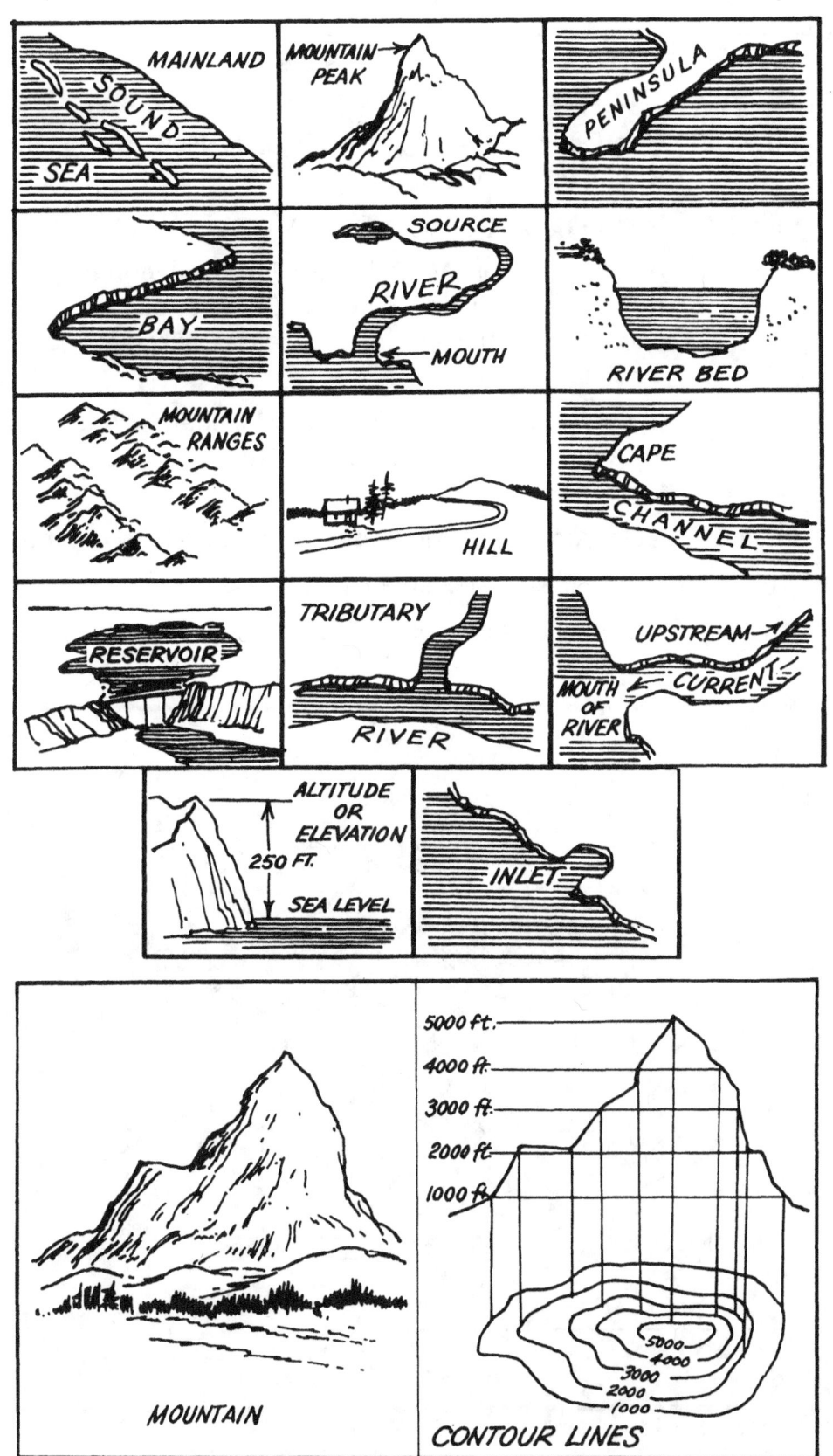

Supplement 2-4-3

Regions of the United States

Region	States
New England	Connecticut, Maine, Massachusetts, New Hampshire, Rhode Island, Vermont
Middle Atlantic	New Jersey, New York, Pennsylvania
Southern	Alabama, Arkansas, Delaware, Florida, Georgia, Kentucky, Louisiana, Maryland, Mississippi, North Carolina, South Carolina, Tennessee, Virginia, West Virginia
Midwestern	Illinois, Indiana, Iowa, Kansas, Michigan, Minnesota, Missouri, Nebraska, North Dakota, Ohio, South Dakota, Wisconsin
Rocky Mountain	Colorado, Idaho, Montana, Nevada, Utah, Wyoming
Southwestern	Arizona, New Mexico, Oklahoma, Texas
Pacific Coast	Alaska, California, Hawaii, Oregon, Washington

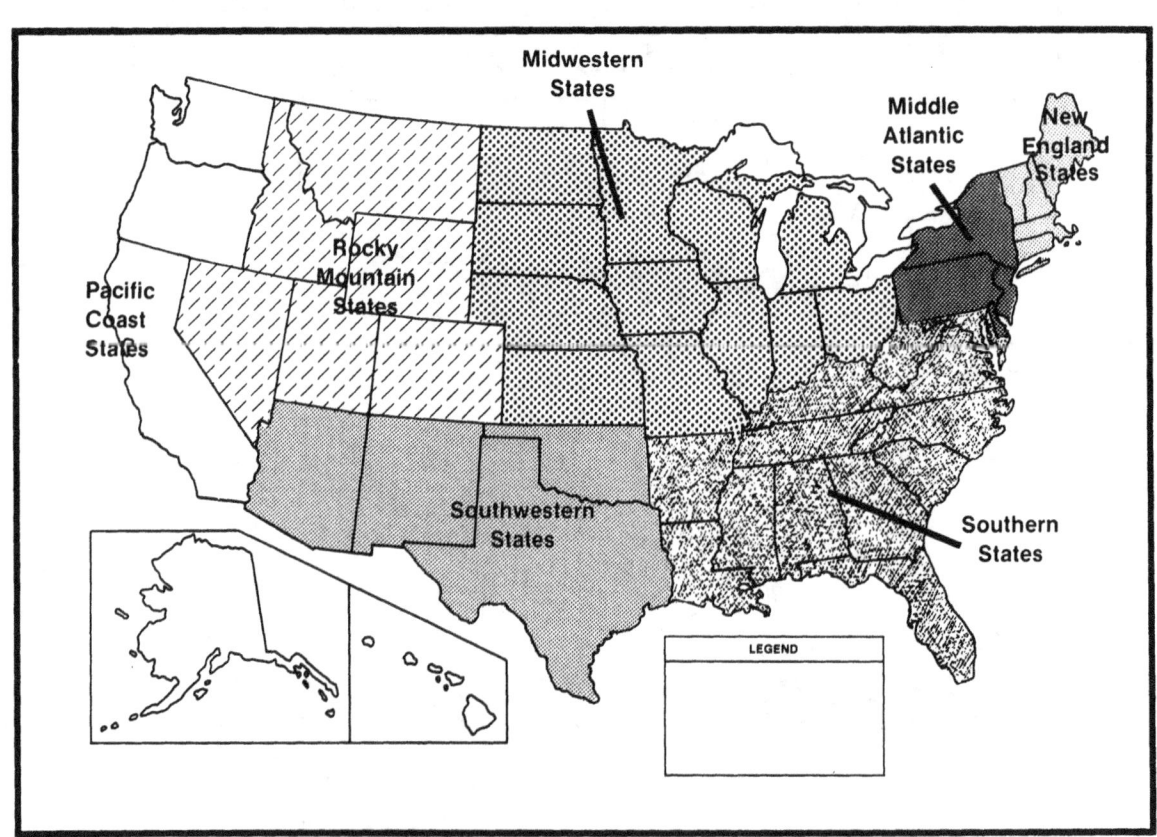

Planning a Highway Route

Imagine that you are a long-distance trucker traveling from coast to coast. Below is a list of eight cities where you will be making deliveries.

Use the map below to plan your route. Fill in the chart with the interstate highways you will be taking from city to city, and indicate the direction you will be traveling to reach each destination. The first stop on your route has been filled in for you.

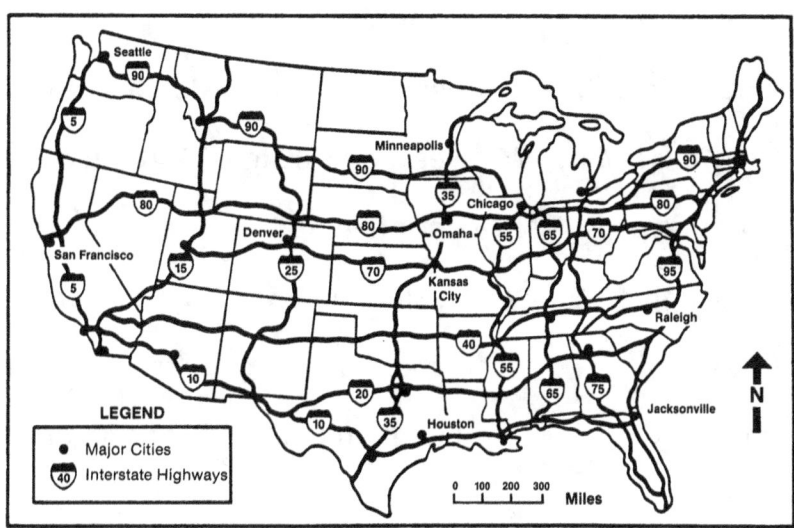

City	Interstate Highway Route and Directions
Jacksonville, FL, to Chicago, IL	west on I-10 to I-75; north on I-75 to I-80; west on I-80
Chicago to Omaha, NE	
Omaha to Seattle, WA	
Seattle to San Francisco, CA	
San Francisco to Kansas City, MO	
Kansas City to Minneapolis, MN	
Minneapolis to Raleigh, NC	
Raleigh to Houston, TX	
Houston to Denver, CO	

Name_____

Planning a Highway Route

Jacksonville, FL, to Chicago, IL	west on I-10 to I-75; north on I-75 to I-80; west on I-80
Chicago to Omaha, NE	west on I-80
Omaha to Seattle, WA	west on I-80 to I-25; north on I-25 to I-90; northwest on I-90
Seattle to San Francisco, CA	south on I-5 to I-80; west on I-80
San Francisco to Kansas City, MO	east on I-80 to I-35; south on I-35
Kansas City to Minneapolis, MN	north in I-35
Minneapolis to Raleigh, NC	south in I-35 to I-40; east on I-40
Raleigh to Houston, TX	south on I-95 to I-10; west on I-10
Houston to Denver, CO	west on I-10 to I-25; north on I-25

Finding the Origin of Geographic Terms and Place Names

Many words used by geographers have interesting histories. Read the examples given below. Then research the origin of the additional names listed. You may use standard or word origin dictionaries and encyclopedias to find your information. You can also look in a geography glossary to discover more names and terms, and explore their origins, too.

glacier: from the Middle French *glace*, meaning ice

globe: from the Latin *globus*, or sphere

1. Australia: _____

2. branch: _____

3. Canary Islands: _____

4. current: _____

5. equinox: _____

6. latitude: _____

7. longitude: _____

8. peninsula: _____

9. port: _____

10. west: _____

Supplement 6-4-3 (answer key)

Finding the Origin of Geographic Terms and Place Names

1. from the Latin *terra australis*, the southern land

2. from the Latin *branca*, meaning paw

3. from the Latin *canis*, or dog, so named because the Spaniards found a race of giant wild dogs there

4. from the Latin *currere*, to run through

5. from the Latin words *aequi* (equal) and *nox* (night)

6. from the Latin *latus*, which implies wide

7. from the Latin *longus*, meaning long

8. from the Latin *paene*, meaning almost and *insula*, meaning island

9. from the Latin *porta*, referring to a gateway

10. akin to Old High German *westar* (to the west) and probably to the Latin *vespera* (evening)

Part FIVE

Resources for Teachers and Students

Part FIVE

Resources for Teachers and Students

The resource lists that follow present only a small sampling of the map and globe materials available to teachers and students. The companies below are among the many suppliers of maps and globes who devote special attention to the school market. You can contact them directly for copies of their catalogs.

George F. Cram Company
P.O. Box 426
Indianapolis, IN 46206

Hammond
515 Valley Street
Maplewood, NJ 07040

Hayes School Publishing
321 Pennwood Avenue
Wilkinsburg, PA 15221

Hubbard Scientific
P.O. Box 104
Northbrook, IL 60062

Modern Educational Systems, Inc.
Publishers Cartographers
Graphic Education
524 E. Jackson Street
Goshen, IN 46526

National Geographic
P.O. Box 1269
Washington, DC 20036

Nystrom
3333 Elston Avenue
Chicago, IL 60618

Rand McNally
P.O. Box 7600
Chicago, IL 60680

World Eagle
64 Washburn Avenue
Wellesley, MA 02181

Many government agencies also provide free or low-cost maps to schools. These include the U.S. Geological Survey, the Census Bureau, and the Government Printing Office.

Print Materials for Teachers

Guidelines for Geographic Education (Washington, DC: Association of American Geographers and National Council for Geographic Education, 1984).
This brief guide presents a rationale for improving geographic education, elucidates the five fundamental themes in geography that have shaped the reforms of the late 1980s, and presents a sequence for geographic education, including instruction in map and globe skills.

Hill, A. David, and Regina McCormick, *Geography: A Resource Book for Secondary Schools* (Santa Barbara, CA: ABC-Clio, 1989).

Despite its title, this book presents much that will be useful to elementary teachers as well, including a thoughtful discussion of the nature of geography and extensive lists of reference and curriculum materials.

Hovinen, Elizabeth L., *Teaching Map and Globe Skills: A Handbook* (Chicago: Rand McNally, 1982).

This comprehensive K-12 guide covers the full range of map and globe skills. An unusual feature is a chapter addressing how map and globe skills can be taught to exceptional students.

Kidron, Michael, and Ronald Segal, *New State of the World Atlas*, 3rd ed. (New York: Simon and Schuster, 1987).

Though too difficult for actual interpretation by elementary students, this unique atlas fascinates students at all levels because of its unusual use of symbols and wide range of topics. It can inspire students to consider design factors when they are making maps themselves. An accompanying activity book, useful for junior high students, is available from CTIR Press.

K-6 Geography: Themes, Key Ideas, and Learning Opportunities (Macomb, IL: Geographic Education National Implementation Project, 1987).

This resource provides several brief teaching ideas for each key idea suggested in the *Guidelines for Geographic Education*. The ideas are organized by grade level and geographic theme; many relate to use of maps and globes.

Maps on File (New York: Facts on File, annual).

More than 375 reproducible maps (updated annually) are in this book. The same publisher also produces *State Maps on File*.

Natoli, Salvatore, J., ed. *Strengthening Geography in the Social Studies* (Washington, DC: National Council for the Social Studies, 1988).

This bulletin focuses on the importance of geography in the social studies curriculum. It provides ideas for classroom activities, as well as guidance in developing a geography program for the entire school.

Rushdoony, Haig A., *The Language of Maps* (Belmont, CA: Pitman, 1983).

A practical book for the classroom, this guide presents 25 lesson plans for teaching map and globe skills. It also includes numerous worksheets, puzzles, and reproducible maps.

Teaching Geography: A Model for Action (Washington, DC: National Geographic Society, 1988).

This three-ring binder summarizes the current reform movement in geography, describing effective classroom activities and listing many sources of assistance to teachers.

Wilfors, J.N., *The Mapmakers* (New York: Alfred A. Knopf, 1981).

A history of cartography, this book is a rich resource for anyone developing lessons on mapmaking.

Print Materials for Students

Bell, Neill, *The Book of Where, or How to Be Naturally Geographic* (Boston, MA: Little, Brown, 1982).

One of the *Brown Paper Book* series, this charming text covers fundamental geographic concepts as well as map and globe skills. The author uses humor and cartoons to interest children and adults alike.

Cheyney, Arnold B., and Donald L. Capone, *The Map Corner* (Glenview, IL: Good Year Books, 1983).

This paperbound book provides lessons, quizzes, and reproducible maps that can be used in map and globe skills learning centers.

Essential Map Skills (Maplewood, NJ: Hammond, 1985).

Basic map skills are covered in this book such as recognizing symbols and legends, calculating distances, using latitude and longitude to locate places, using a map index, and interpreting map projections. Many colorful maps are provided.

Job, Kenneth, and Lois Weiser Wolff, *Skills for Understanding Maps and Globes* (Englewood Cliffs, NJ: Prentice Hall/Allyn and Bacon, 1986).

This workbook presents 19 lessons focusing on interpreting various kinds of maps. The lessons are designed for use with junior high school students.

Knaup, Marianne, *Map Skills Series* (St. Louis, MO: Milliken, 1986-1988).

This series of five worksheet/transparency books profiles various regions of the United States. As students learn about the regions, they also practice their map-reading skills.

Map and Globe Skills (Culver City, CA: Social Studies School Service).

All aspects of map reading are covered in this two-part program. The materials, which are designed for use in learning centers, include large visual displays and accompanying worksheets.

Map Skills for Today (Columbus, OH: Weekly Reader, 1986).

This workbook series offers a book for each grade level, 1-6. The workbooks present a sequential approach to map skill development correlated with the "expanding horizons" approach generally used in elementary social studies. Weekly Reader also produces a map skills poster kit for grades 4-6.

Maptime . . . USA (Carthage, IL: Good Apple, 1982).

A handy resource, this book provides reproducible map activities that cover skills related to location/direction, symbols/legends, distance/scale, and the use of road maps.

Nero, Ann Briscoe, *Essential Skills in Geography: A Workbook of Basic Geographic Concepts* (Chicago: Rand McNally, 1987).

Designed for students in grades 3-6, this paperbound text covers not only use of maps, but also skills needed to read and interpret many kinds of graphs.

Audiovisual Materials

Building Map and Globe Skills (Chicago: Encyclopaedia Britannica, 1980).
This five-part filmstrip program focuses on location, direction, distance/scale, map symbols, elevation, movement of the earth, and cartography.

How to Use Maps and Globes (Bohemia, NY: Charles Clark Company, 1987).
A 49-minute video, this program focuses on what maps and globes are, how they are used, and the global grid system. Uses of latitude and longitude are highlighted.

Introducing Maps and Globes Series (New York: BFA Educational Media, 1978).
This four-part video program introduces what maps and globes are and how they are used. Students then learn about use of symbols, latitude and longitude, earth movements, and different kinds of maps.

Map Competency (New York: Globe Filmstrips, 1979).
Consisting of six filmstrips, this program helps slow and average junior high students understand fundamental map-reading skills. Special attention is given to how physical features are shown on maps.

Maps and Globes Skills Box (Mahway, NJ: Troll).
This kit uses eight cassettes and 40 worksheets to teach basic map skills, including finding physical features, using latitude and longitude, and interpreting maps.

Maps and Globes: An Introduction (Studio City, CA: FilmFair Communications, 1981).
This 17-minute film describes the various kinds of information maps provide. Such map-reading skills as interpreting symbols and calculating distances using the scale are covered.

Maps and What We Learn From Them (Washington, DC: National Geographic Society, 1987).
Intended for students in grades 3-6, this three-filmstrip program looks at the uses of maps and how they are made. Early and modern maps are compared.

Using Maps, Globes, Graphs, Tables, and Charts (Niles, IL: United Learning, 1984).
This is a five-part filmstrip series. One filmstrip is devoted to each of the items listed in the title.

Where and Why (Chicago: Nystrom, 1980).

Included in this comprehensive program are 24 cassettes, numerous worksheets, 150 study prints, and a teacher's guide. The cassettes provide directions to complete the 24 lessons, which focus on basic map and globe skills.

Computer Software

Discovery (Chicago: Nystrom).

This four-part tutorial program, available for Apple equipment only, covers the basics of map reading (symbols, direction, distance, and grids). It then moves on to the skills of locating, using latitude and longitude, and interpreting thematic maps.

Language of Maps, The (Garden City, NY: Focus Media).

Another tutorial program for Apple computers, *The Language of Maps* presents a great deal of geographic information (e.g., continents, water bodies, highlands/lowlands) as it teaches students to use the map legend, compass rose, grid system, and scale.

Latitude and Longitude (Wichita, KS: Learning Arts).

As the title suggests, this tutorial program for Apple computers focuses on latitude and longitude. Students are taught not only how to use the global grid system, but also why understanding its use is important.

Map Pack (Geneva, IL: Penguin Software).

This Apple package generates maps of the 50 states, the Canadian provinces, and the continents. Students can add to the maps.

Map Reading: An Introduction to Direction and Distance (Logan, IA: Perfection Form).

Available only for Apple computers, this tutorial program focuses on the two concepts mentioned in the title—teaching students to read a compass and to calculate distances.

Maps (Minneapolis, MN: JMH Software).

This drill-and-practice program is designed for Commodore or Atari computers. The three-part program focuses on direction, latitude, and longitude.

Maps and Globes (Edison, NJ: Micro-Ed).

Basic map skills are covered in this drill-and-practice program. It can only be used with Apple, Commodore, and Radio Shack equipment.

Maps and Globes Micro (Mahwah, NJ: Troll).

The first part of this tutorial package focuses on basic map skills but also teaches students about landforms and bodies of water. The second part covers latitude. The package is available for Apple systems only.

Unlocking the Map Code (Chicago: Rand McNally).

This package, which is available for Apple and Atari computers, covers land and water forms, interpreting color and map symbols, direction, location, scale, and time. Students apply what they have learned in developing a simulated flight plan.

Multimedia Kits

Essential Skills in Social Studies: Levels A-C (Chicago: Rand McNally, 1988).

Each of the three kits in this series for the primary grades contains a wide variety of instructional materials, including books, copy masters, laminated desk maps (multiple copies of several different maps in each kit), and posters. The program, which puts strong emphasis on map and globe skills, is correlated with the "expanding horizons" approach.

Flight: A Simulation of a Cross-Continent Air Race (Lakeside, CA: Interact).

Developed for students in grades 4-8, this simulation program builds map-reading and decision-making skills. Teams of students plan travel routes, keep daily diaries, and solve a variety of problems as they race across the United States. A computer supplement is available.

Geodyssey (Culver City, CA: Social Studies School Service, 1987).

Created for students in third grade and up, this game involves building a world map with two-inch plastic tiles.

Graphic Learning Social Studies (Tallahassee, FL: Graphic Learning Corporation, 1986).

This series includes programs for grades 1-6 in studies of the United States and Latin America. It emphasizes map skills and is correlated with the "expanding horizons" approach. Each program in the series includes copy masters, multiple laminated desk maps, marking pens, and a teacher's guide; some also include student books.

Hands-On Geography (Chicago: Nystrom, 1986).

Consisting of seven separate programs, this series covers the world, the United States, Europe, Asia, Africa, North America, and South America. Map skills are emphasized. Each program includes multiple copies of desk maps, a teacher's guide, copy masters, and marking pens.

Map Projection Model (Northbrook, IL: Hubbard Scientific).

This unique kit of materials allows students to experience how flat maps are made to represent the curved surface of the earth. Several map projections are included to facilitate comparisons.

Where in the World? (Ann Arbor, MI: Aristoplay, 1986).

For students in grades 3 and up, this game consists of six region boards, 174 country cards, and five wild cards. Map skills are reinforced as students answer questions to progress through the game.

World Traveler (Culver City, CA: Social Studies School Service).

Designed for student in grades 2-6, this game requires players to answer geography questions as they move their markers across a world map. A *U.S. Traveler* game is also available.

Part SIX

Appendix

Part SIX

Appendix

Fascinating Facts about Maps and Globes

1. The word *map* comes from the Semitic language through Medieval Latin. At one time the word meant *cloth* or *napkin*. Yet maps are much older than symbols appearing on fabric. Some of the earliest maps were of rock, wood, clay, marble, animal hide, and papyrus. Today, raised relief maps are made of plastic.

2. The oldest known map of any kind was made about 3800 B.C. It is a small clay tablet showing the Euphrates River flowing through northern Mesopotamia (Iraq).

3. The Egyptians made maps as early as 1300 B.C., but the Greeks drew the first accurate maps. In 500 B.C. they became one of the first groups to realize that the earth is round. They designed the first projection map and developed a longitude and latitude system.

4. The Roman emperor Augustus had a map made of the entire Roman Empire. Completing the map took almost twenty years. Copies of it were made on papyrus rolls (22 feet long and 13 inches wide) folded in a portfolio.

5. The Romans used maps to establish taxation systems and to plan military strategies. By A.D. 350 they had mapped a network of roads from southern England to the Ganges Valley in India.

6. Ptolemy, an Egyptian scholar, constructed a map of the known world about A.D. 150. It represented a spherical surface on a plane. He also compiled a list of 8,000 place names and produced an atlas of sectional maps. Ptolemy determined his latitudes astronomically and made attempts to ascertain longitude by timing eclipses. Until the 1500s, maps were based on the writings of Ptolemy.

7. Centuries ago, Bedouins and Polynesians made travel maps. Eskimos, without the benefit of surveying instruments, carved maps of northern Canada on bone and wood. They resemble current charts of those regions made by recent surveyors and cartographers.

8. The Chinese are credited with making the first printed map around A.D. 1150. Some claim the earliest printed map of the world appears in Isadore of Seville's *Etymologiarum* of 1472.

9. Religion had considerable influence on mapmakers. Because the Bible speaks of the four corners of the earth, as well as the circuit of the earth, early mapmakers drew square maps as well as oval ones. Jerusalem, the Holy City, was usually located at the center of these maps.

10. To early mapmakers of southern and central Europe, east was more important than north. The sun rose in the east, and the Holy Land lay in that direction. Also, the great wealth that eventually lured Columbus and other explorers was in the East, or Orient. Thus, when the map was placed with its top to the east, it was said to be oriented. Placing north, instead of east, at the top of a map is a relatively modern practice.

11. As people began long pilgrimages, maps were gradually improved. During the Crusades, men began to use the compass in sailing, and maps became more accurate.

12. The great discoveries made by Christopher Columbus, John Cabot, and others during the fifteenth and sixteenth centuries were laid down on compass charts developed in the Mediterranean area. Chart-making flourished in the Italian port cities of Genoa, Ancona, and Venice and in Lisbon, Portugal.

13. Christopher Columbus was a very skilled mapmaker.

14. In 1492 Martin Behaim (1459-1507), a German merchant-navigator, made the oldest existing globe.

15. The first map to use the name "America" was made in 1507 by Martin Waldseemuller (1470-1518), a German mapmaker. It measured about 4½' x 8'.

16. Gerhard Mercator (1512-1594) was a famous Flemish mapmaker in the 1500s. He produced some of the best maps and globes of his time and developed a map projection for sailors.

17. Abraham Ortelius (1527-1598) produced the first modern atlas in 1570.

18. The first map of Virginia was drawn by the famous adventurer, John Smith (1580-1631).

19. Before launching his military and political career, George Washington was a surveyor and mapmaker.

20. American explorers Meriwether Lewis, William Clark, Zebulon Pike, Jedediah Smith, and John C. Frémont all provided information about the West. Much of it was incorporated in the maps of the 1800s, which were used by pioneers and gold-seekers in their journeys west.

Making Comparisons: Highest, Biggest, Longest, Widest
Continental Statistics

Name of Continent	Area in sq. mi.	% of Earth's Surface	Estimated Population (1988)	% of World's Population	Highest Point	Lowest Point
Asia	17,300,000	29.9	3,031,100,000	60.0	Mt. Everest, Nepal-Tibet, 29,028 ft.	Dead Sea, Israel-Jordan, 1,312 ft. below sea level
Africa	11,700,000	20.2	615,300,000	12.2	Mt. Kilimanjaro, Tanzania, 19,340 ft.	Lake Assal, Djibouti, 512 ft. below sea level
North America	9,400,000	16.3	413,000,000	8.2	Mt. McKinley, Alaska (USA), 20,320 ft.	Death Valley, California (USA), 282 ft. below sea level
South America	6,900,000	11.9	282, 200,000	5.6	Mt. Aconcagua, Argentina, 22,834 ft.	Valdes Peninsula, Argentina, 131 ft. below sea level
Europe	3,800,000	6.6	684,800,000	13.5	Mt. Elbrus, Soviet Union, 18,510 ft.	Caspian Sea, Soviet Union-Iran, 92 ft. below sea level
* Australia	3,300,000	5.7	25,500,000	0.5	Mt. Kosciusko, Australia, 7,310 ft.	Lake Eyre, Australia, 52 ft. below sea level
Antarctica	5,400,000	9.3	———	——	Vinson Massif, Antarctica, 16,864 ft.	Unknown

Estimated world population in 1988: 5,128,000,000

* Population includes people of New Zealand and other islands in Oceania. The highest point in this region is Mt. Cook in New Zealand at 12,349 ft.

The Oceans of the World

Ocean	Area in sq. mi.	Average depth in ft.
Pacific	64,186,300	12,925
Atlantic	33,420,000	11,730
Indian	28,350,500	12,598
Arctic	5,105,700	3,407

Five Largest Deserts of the World

Desert	Location	Area in sq. mi.
Sahara	North Africa	3,500,000
Gobi	Mongolia, China	500,000
Libyan	Libya, Egypt, Sudan	450,000
Kalahari	Southern Africa	225,000
Gibson	Australia	125,000

Ten Largest Countries of the World in Area

Country	Area in sq. mi.
Soviet Union	8,649,496
Canada	3,851,790
China	3,705,390
United States	3,618,770
Brazil	3,286,470
Australia	2,966,200
India	1,266,595
Argentina	1,065,189
Sudan	966,757
Algeria	918,497

Ten Most Populated Urban Areas of the World

City	Population
Tokyo-Yokohama, Japan	25,434,000
Mexico City, Mexico	16,901,000
São Paulo, Brazil	14,911,000
New York City, USA	14,598,000
Seoul, Korea	13,665,000
Osaka-Kobe-Kyoto, Japan	13,562,000
Buenos Aires, Argentina	10,750,000
Calcutta, India	10,462,000
Bombay, India	10,137,000
Rio de Janeiro, Brazil	10,116,000

Natural Phenomena

1. Greatest earthquake: Concepción, Chile, May 22, 1960, assessed at 9.5 on the Richter scale

2. Greatest volcanic eruption: Tambora on the island of Sumbawa, Indonesia, April 5-7, 1815, estimated at 36.4 cubic miles

3. Tallest geyser: Waimangu in New Zealand, erupted to a height of more than 1,500 feet in 1904

4. Point farthest from land: 48° south latitude, 125° west longitude

5. Point farthest from the sea: Northern Xinjiang, China, 1,500 miles from the open sea in any direction

6. Coldest sea temperature: White Sea, Soviet Union, 28.5°F average surface temperature

7. Warmest sea temperature: Persian Gulf, 96°F average surface temperature

8. Largest gulf: Gulf of Mexico, 580,000 square miles

9. Largest bay: Hudson Bay, Canada, 317,500 square miles

10. Longest strait: Tatar Strait, between Sakhalin Island and the USSR mainland, 497 miles long

11. Largest peninsula: Arabia, 1,250,000 square miles

12. Remotest island: Bouvetöen (54°S, 3°E) in South Atlantic, 1,050 miles from the nearest land

13. Greatest archipelago: 3,500-mile-long crescent of over 13,000 islands that form Indonesia

14. Deepest lake: Lake Baykal, Siberia, USSR, 4,872 feet below sea level

15. Highest waterfall: Angel Falls, Venezuela, with a total drop of 3,212 feet

16. Longest fjord: Nordvest Fjord, Greenland, extends 195 miles from the sea

17. Largest swamp: Pripya in the Soviet Union, 18,125 square miles

18. Longest cave: Mammoth Cave, Kentucky, total of 184.64 miles

19. Largest gorge: Grand Canyon, Arizona, total of 217 miles

20. Highest temperature recorded: Al Aziziyah, Libya, September 13, 1922, 136.4°F

21. Lowest temperature recorded: Oymyakon, Siberia, USSR, in 1964, –96°F

22. Greatest snowstorm recorded: Shasta Ski Bowl, California, 189 inches

23. Greatest rainfall in a 24-hour period: Cilaos, La Reunion, Indian Ocean, March 15-16, 1952, 73.62 inches

24. Wettest place (annual mean): Mt. Waialeale, Kauai, Hawaii, 451 inches of rain

25. Driest place (annual mean): Calama, Atacama Desert, Chile, no rainfall

26. Windiest place: The Commonwealth Bay, George V coast, Antarctica, where gales reach 200 miles per hour

Five Longest Rivers of the World

River	Location	Approx. length in mi.
Nile	Africa	4,160
Amazon	South America	4,000
Mississippi-Missouri	North America	3,880
Yangtze	Asia	3,602
Ob	Asia	3,362

Five Longest Ship Canals of the World

Canal	Location	Length in mi.
Baltic-White Sea	Soviet Union	141
Suez	Egypt	101
Albert	Belgium	81
Moscow-Volga	Soviet Union	80
Kiel	Germany	61

Five Largest Islands of the World

Island	Location	Area in sq. mi.
Greenland	North America	840,000
New Guinea	Oceania	306,000
Borneo	Asia	286,914
Madagascar	Africa	226,658
Baffin	North America	195,928

Five Largest Natural Lakes of the World

Lake	Location	Area in sq. mi.
Caspian	Asia	143,244
Lake Superior	North America	31,700
Lake Victoria	Africa	26,828
Aral'skoye	Asia	24,904
Lake Huron	North America	23,000

Map Masters

World Map with Latitude and Longitude Grids 182-183

Map of the World: Major Mountain Ranges and Rivers 184

Map of Africa 185

Map of Antarctica 186

Map of Asia 187

Map of Australia, New Zealand, Oceania 188

Map of Europe 189

Map of North America 190

Map of South America 191

Map of the United States: Major Mountain Ranges and Rivers 192

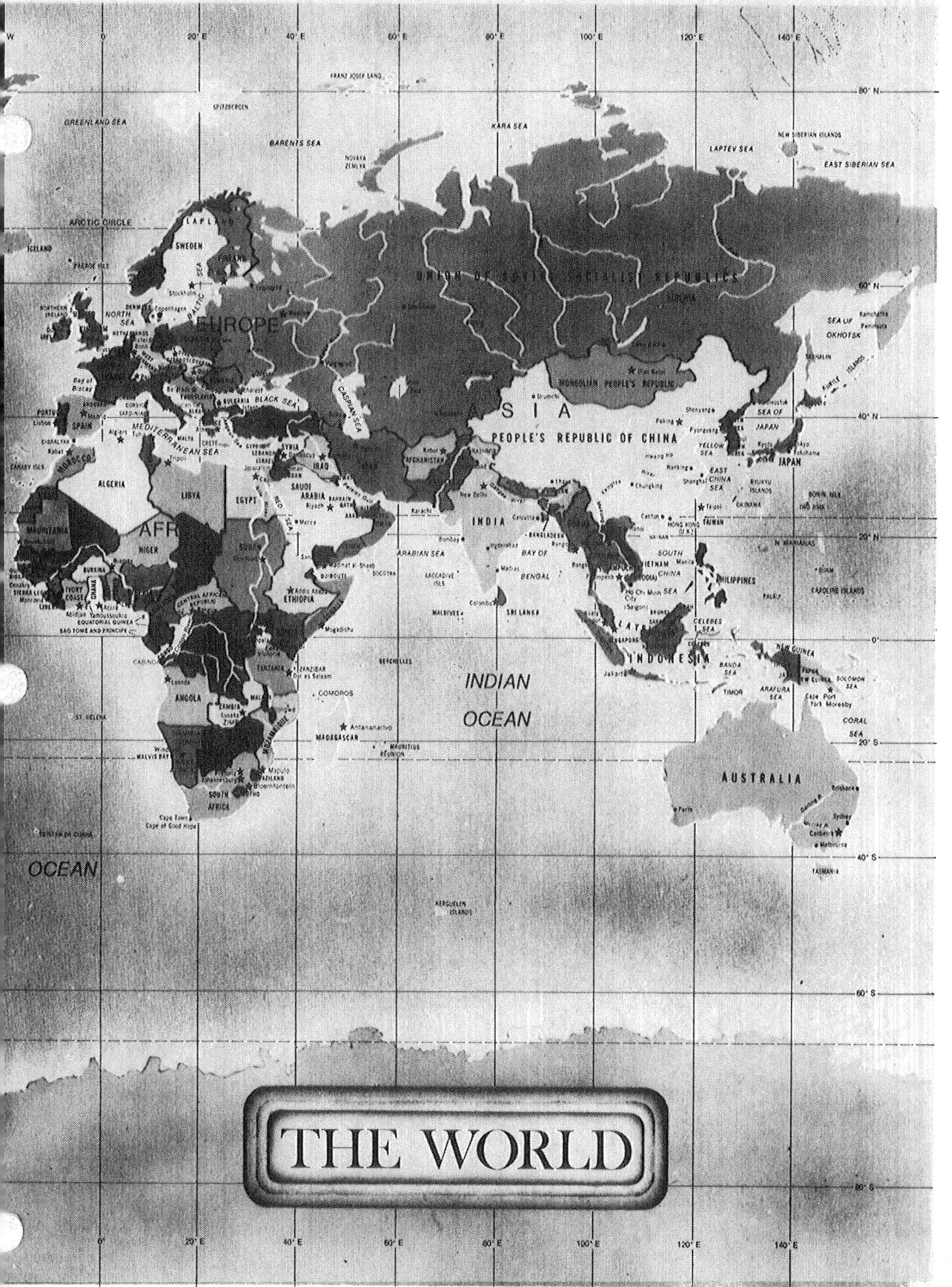

THE WORLD

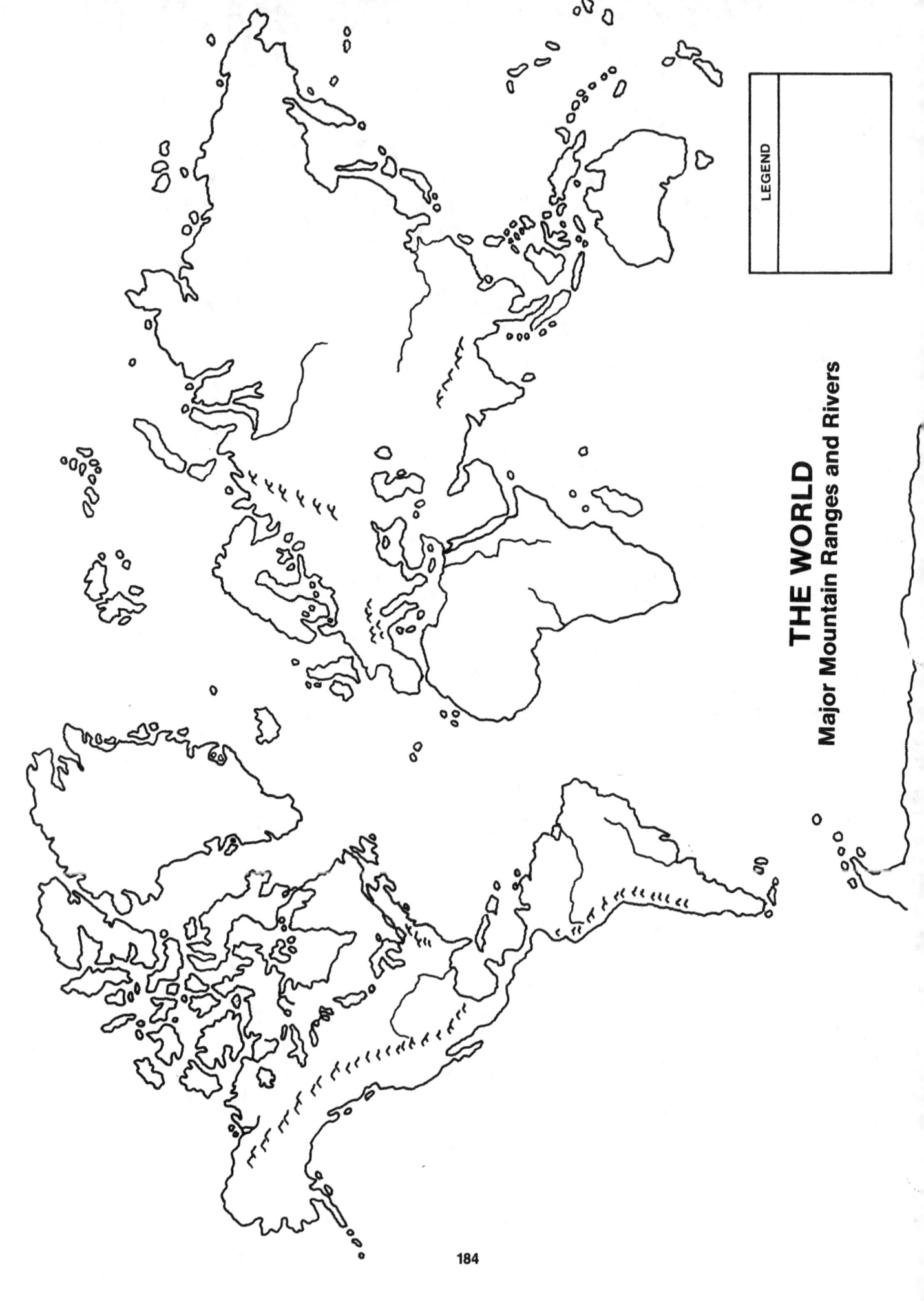

THE WORLD
Major Mountain Ranges and Rivers

LEGEND

Africa

LEGEND

Antarctica

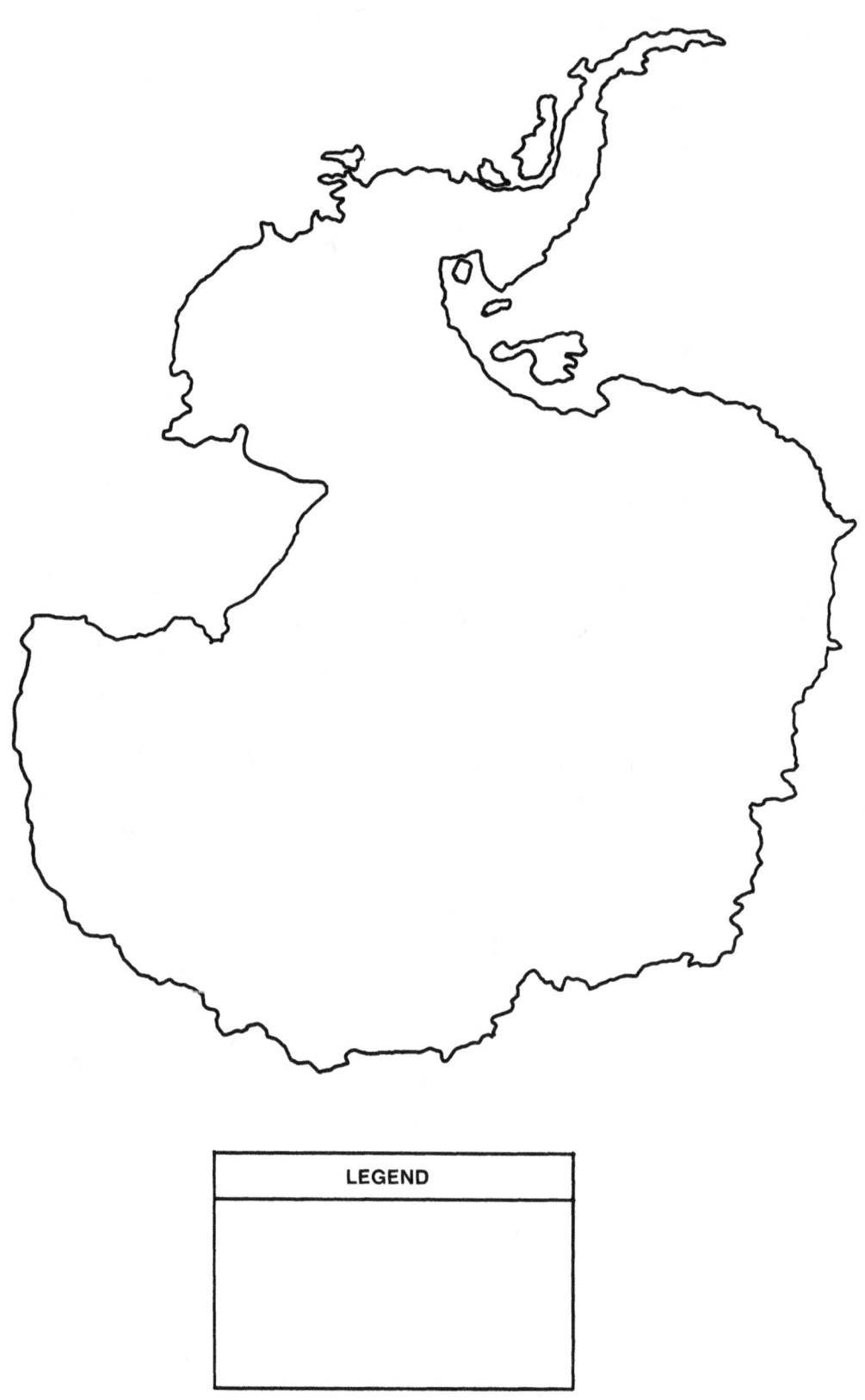

LEGEND

Asia

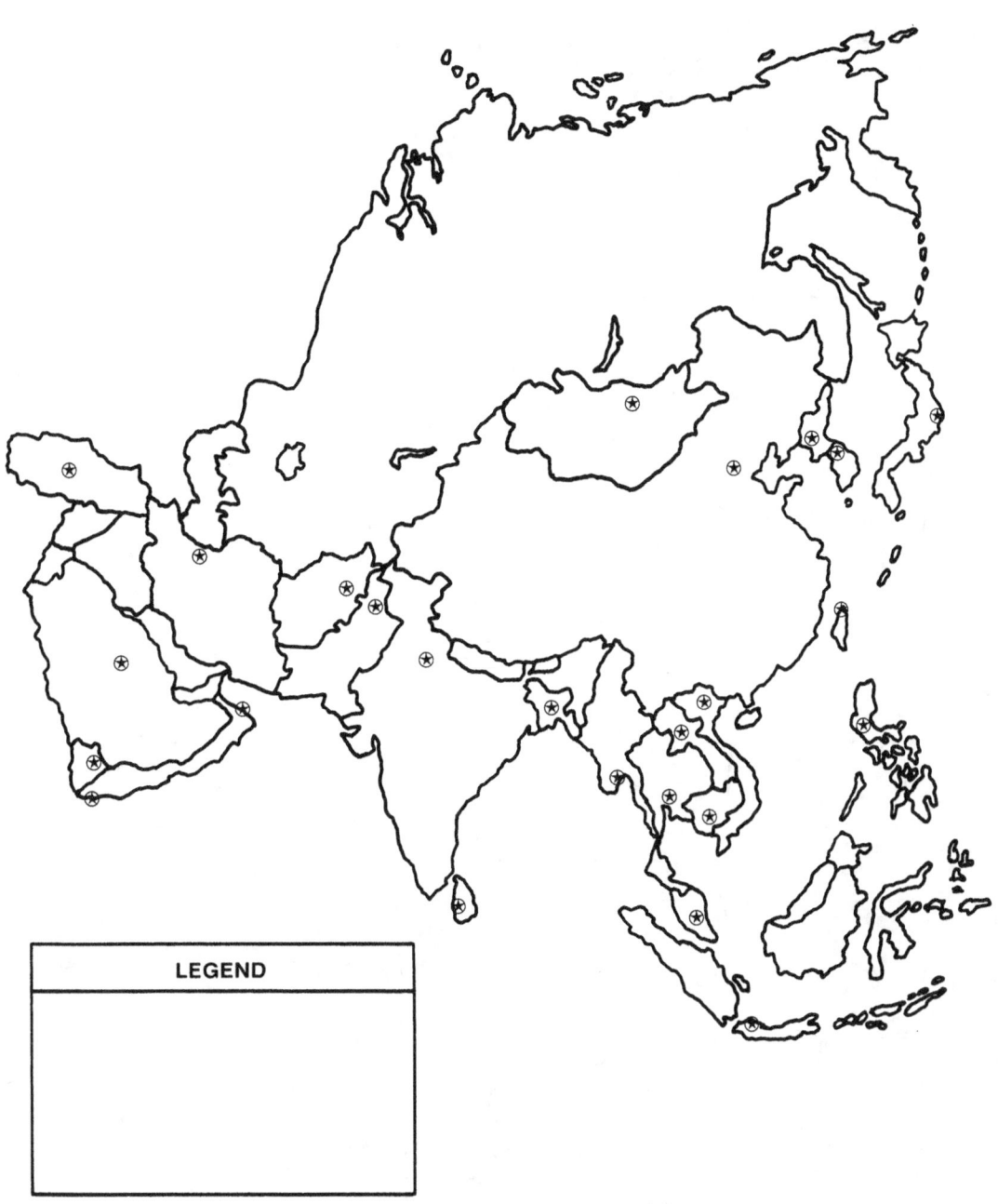

LEGEND

Australia, New Zealand, Oceania

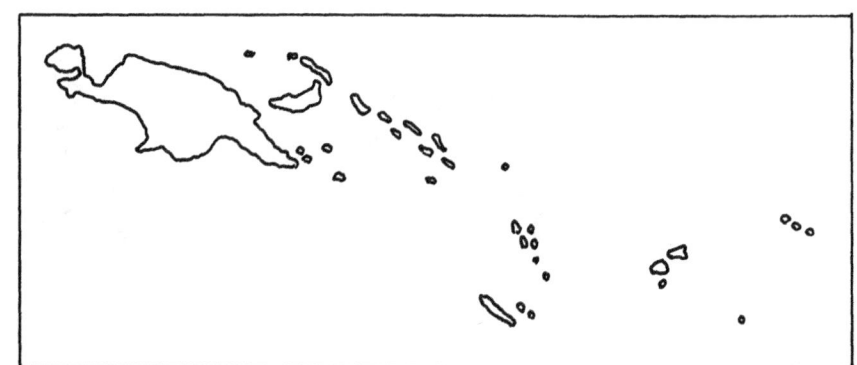

LEGEND

Europe

LEGEND

North America

LEGEND

South America

LEGEND

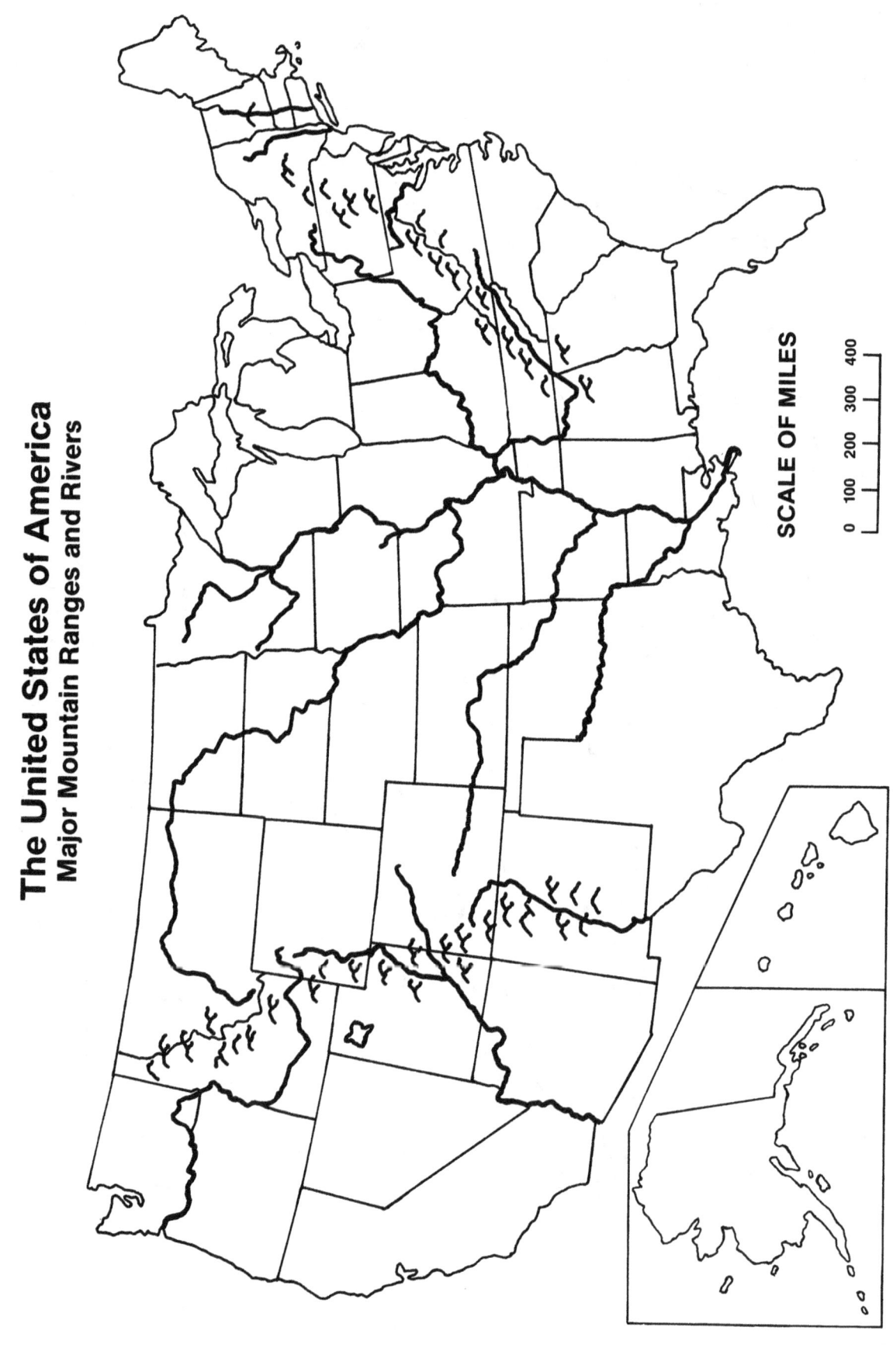

The United States of America
Major Mountain Ranges and Rivers

SCALE OF MILES

0   100   200   300   400

Glossary

altitude The height of an object above a given level.

Antarctic Circle The parallel of latitude 66½° south of the equator. At the winter solstice, all parts of the earth south of this line remain light for 24 hours.

Arctic Circle The parallel of latitude 66½° north of the equator. At the summer solstice, all parts of the earth north of this line remain light for 24 hours.

atlas A book presenting a collection of maps.

axis An imaginary rod that runs through the earth, from pole to pole. The earth rotates around this axis.

boundary A line that marks the outer edge of a political feature such as a city, state, or nation. Boundaries can be imaginary human-drawn lines or natural features such as rivers.

cardinal directions The four main points of the compass: north, south, east, west.

cartogram A special map in which sizes of countries are based on some measure other than area (e.g., population).

cartographer A person who makes maps or charts.

compass rose A designed symbol on a map that shows the four cardinal directions.

continent A large area of land on earth. The seven continents are North America, South America, Australia, Antarctica, Africa, Europe, and Asia. Europe and Asia are sometimes considered a single continent, Eurasia.

contour lines Lines on a map or chart connecting all points in a region that have the same elevation (or depth).

data Pieces of factual or assumed information from which conclusions can be drawn and inferences can be made. A great deal of data can be readily plotted on maps.

degree A unit of measure equal to 1/360th of the circumference of a circle. Latitude and longitude are measured in degrees.

distortion Inaccuracies in size, shape, or direction caused by showing the spherical earth on a flat map.

distribution The arrangement of objects over an area. Distribution is often shown on maps.

down The direction toward the center of the earth. Down is not the same as south on a map.

east One of the cardinal directions. It is the direction toward which the earth turns and in which the sun rises. East is at 90° on the compass and to the right when facing north.

elevation The height of land above sea level.

elevation map A map that shows how far above or below sea level land is. This information is shown using contour lines or colors.

environmental map A map that shows the kind of surroundings in an area. Environmental maps include data on climate, vegetation, and landforms.

equator An imaginary line that circles the middle of the earth, equally distant from the North and South poles. The equator divides the earth's surface into the Northern and Southern hemispheres. The equator is 0° latitude.

equinox The two times of the year when the sun crosses the equator, causing day and night to be of equal length everywhere on the planet. One equinox occurs on March 21 or 22 and the other on September 21 or 23.

gazetteer A geographical dictionary, listing places in alphabetical order and their description.

general-reference map A map presenting several kinds of information, including physical and political features.

globe A model of the earth, showing land and water areas. Globes are usually attached to a stand to show the tilt of the earth; many globes can spin around an axis.

great circle route A course plotted along any circle around the earth's circumference. Such a course, used by airplanes, is the shortest route between any two points on the earth's surface.

grid A system of horizontal and vertical lines used to locate places on a map. Latitude and longitude are one example of a grid system.

grid point The place where two lines in a grid system cross each other.

grid square A square created by four lines in a grid system; in alphanumeric grids, squares (rather than points) are used to locate places on a map.

hemisphere Half a sphere. The earth is divided by the equator into the Northern and Southern hemispheres. The prime meridian and 180° longitude divide the earth into the Eastern and Western hemispheres.

high latitudes Areas north of the Arctic Circle and south of the Antarctic Circle; they have cold, dry climates.

inclination The tilting of the earth's axis.

intermediate directions The directions between the cardinal directions— northwest, northeast, southwest, and southeast.

international color scheme A series of graduated colors used to indicate bodies of water, topography, land elevations, and sea depths. Shades of blue, white, green, yellow, amber, and orange are normally used.

International Date Line The line going from pole to pole near the 180° longitude. Crossing the line going east, a traveler gains a calendar day. Going west across the line, a traveler loses a day.

interrupted projection A map projection that splits the oceans in order to show land areas more accurately.

isobar A line on a map connecting places having equal atmospheric pressure.

isotherm A line on a map connecting places having equal average temperatures.

key A list explaining what the symbols on a map stand for.

latitude The distance north or south of the equator. Measured in degrees, minutes, and seconds. The highest latitude is 90° south or north—the two poles. Lines of latitude are also called parallels.

legend Another name for the map key.

location The position of a point on the earth's surface. Location can be described most precisely using latitude and longitude.

longitude The distance east or west of the prime meridian. Longitude is measured in degrees, minutes, and seconds. The highest degree of longitude is 180°, the line near the International Date Line. Lines of longitude are also called meridians.

low latitudes Area between the Tropic of Cancer and the Tropic of Capricorn, including the equator. The climate here is hot year round, but can be either wet or dry.

map A flat drawing of the earth's surface, drawn to scale and using symbols to convey information about the earth.

map index An alphabetical listing of places shown on a map, with references to the grid point or grid square where each place can be found.

Mercator projection A map projection in which all lines of latitude and longitude are shown as being straight. This projection is useful to sailors because it shows direction accurately. However, shape and size are distorted, especially near the poles.

meridians Lines of longitude.

middle latitudes The area between the tropics and the Arctic and Antarctic circles. Climate in the middle latitudes reflects seasonal changes.

north One of the cardinal directions. North is toward the North Pole. On a compass, north is at 0° or 360°, directly opposite south.

ocean The large body or bodies of saltwater that cover more than half the earth's surface. Some geographers believe there is only one large ocean while others identify four oceans: Atlantic, Pacific, Indian, and Arctic.

parallels Lines of latitude. Called that because they are parallel to the equator; that is, at every point on a given line of latitude, the equator is the same distance away.

physical feature Natural feature of the earth's surface, such as a hill, mountain, or river.

place A location having special characteristics that make it different from other places.

polar projection map A projection that shows the earth with one of the poles at the center. On this type of map, the land areas around the equator are distorted. A polar projection is useful to pilots in determining shortest routes from place to place.

political feature Objects or features created by humans such as roads, cities, or buildings.

prime meridian The meridian from which longitude is measured both east and west. It is designated as 0° longitude. The prime meridian passes through Greenwich, England.

projections The ways in which cartographers transform locations on the spherical earth to locations on a flat map. Every projection involves some distortion.

region An area of land in which all places are similar in one or more ways, giving the area a set of common characteristics. Regions can be based on similarities in natural features (for example, a mountainous region) or political or cultural features (for example, a region in which the people share a common religion).

relief map A map showing the elevation and surface irregularities of the land above sea level and the depth of the sea.

revolution The movement of the earth around the sun every 365¼ days.

Robinson projection A projection that distorts size, shape, and direction somewhat but does not distort any one of these as noticeably as most other projections.

rotation The spinning of the earth on its axis every 24 hours.

satellite photo A photograph taken from above earth by a special camera in a satellite. Satellite photos have been useful in mapping unexplored areas of the earth. Meteorologists superimpose maps over satellite photos to create special satellite photo maps.

scale Way of showing how much distance on earth is represented by a given distance on a map.

solstice The two times of the year when the earth is farthest from the sun. The solstices are in June and December.

south One of the cardinal directions. South is toward the south pole. On the compass, south is at 180°, directly opposite north.

special-purpose map A map drawn for a particular reason and usually showing only one kind of information (for example, natural resources).

symbol A line, color, shape, or picture used to represent a real object.

thematic map A special-purpose map.

time zone One of the 24 divisions of the earth. Each is approximately 15° of longitude; the time in each zone is one hour earlier than in the time zone to the east and one hour later than in the time zone to the west.

topographical map A map showing the elevation and natural features of an area.

Tropic of Cancer A parallel on the earth about 23½° north of the equator. It marks the latitude farthest north that receives the vertical (direct) rays of the sun. The sun is directly over the Tropic of Cancer on June 21 or 22 (the summer solstice).

Tropic of Capricorn A parallel on the earth about 23½° south of the equator. It marks the latitude farthest south that receives the vertical (direct) rays of the sun. The sun is directly over the Tropic of Capricorn on December 21 or 22 (the winter solstice).

up The direction away from the center of the earth. Up is not the same as north on a map.

west One of the cardinal directions. The earth spins away from the west, making it the direction in which the sun sets. West is at 270° on the compass and on the left when one is facing north.

Index

American Indian tribes map, 81
audiovisual materials, 168-169

barometric tendency, symbols, 145
basic concepts
    direction and location, 19-25
    grids, 30-36
    for intermediate grades, 8-10
    interpreting maps, 37-43
    for junior high grades, 10-11
    making maps, 43-50
    for primary grades, 7-8
    scale and distance, 26-29
    symbols and legends, 13-19
    tips for teachers, 11-12
Beaufort scale of winds, 146
buildings, tallest in U.S., 140-142

capitals, coordinates for, 128-135
cartographer's corner, 12
cities
    matching sister cities, 119-121
    same names activity, 116-121
community maps, 66
compass rose, 21, 107-108
computer software, 169-170
continental statistics, 175-178
cross-curriculum activities
    current world events, 86-89
    home, school, neighborhood
      maps, 54, 65-67
    state maps, 67-73
    U.S. and world maps, 73-85

delivery route maps, 60-61, 62,
    161-162
direction and location
    compass rose, 21, 107-108
    earth's movement, 19-21
    location coordinates and capitals,
      128-135
    location coordinates and foods,
      136-138
    location coordinates and world
      map, 128-130
    sun observation, 56-58
distance. See scale and distance

earth
    globe as model of, 13-14
    movement of, 19-21
economic activities map, 75-76
European settlement, in U.S., 82-83

flight plan, 139
floor plans, 51-53, 149
foods, and locations coordinates,
    136-138

general-reference maps, 37-38
    and location, direction, 38-39
geography
    and American Revolution, 82-83
    continental statistics, 175-178
    flash cards for, 79
    language of maps, 73-74
    natural features, 153-159
    origin of terms, place names, 80,
      163-164
    and world events, 86
global grid quiz, 88
globe
    distances on, 29-30
    grid, 33
    model of earth, 13-14
    papier-mâché, 46
glossary, 193-197
goals, of program, 7-11
grids, 30-36, 126-127
    additional activities, 35
    global, 32-35
    latitude, 36
    longitude, and global time, 33-35
    using, 31-32, 33-35

what is a grid, 30-31

hidden words activity, 112
highway map symbols, 98-102
holidays, and maps, 62
home, floor plan of, 51-53, 149

innovative maps, ingredients for, 46-50
inset maps, 28-29
intermediate grades, basic concepts for, 8-10

junior high grades, basic concepts for, 10-11

latitude
  on global grid, 32-37
  locating coordinates, 128-135
legends. *See* direction and location
longitude, 32-35, 128-135

map search, 113-115
maps, interpreting
  different kinds of, 37-38
  fact or fiction, 38-39, 141
  identifying region, 41-42
  jeopardy map board, 42
  making inferences, 41-42, 148
  mystery location, 41-43
  road maps, 39-40
  travelog, 43, 152
  weather maps, 40-41, 55-58, 144-147
maps, making
  delivery route, 60-61, 62, 161-162
  floor plan, 51-53, 149
  home, 51-53
  innovative maps, 46-50
  neighborhood, 58-59
  papier-mâché globes, 46
  playground, 44-45
  safety symbols, 55-56
  seating charts, 53
  state, 69-73
  sun observations, 56-58

  terrain, 60
  three-dimensional relief, 45-46
  weather and seasonal, 54-55
maps, state
  advertising state, 72
  environmental maps, 72-73
  history maps, 70-71
  mileage chart, 68-69
  news coverage, 86-89
  place names, 67-68
  relief maps, 72
  scrambled states, 103-105
  state fair maps, 71-72
  thematic map, 69-70
maps, U.S. history
  American Indian tribes, 80-81
  American Revolution, 82-83
  conserving wilderness, 86
  dust bowl, 86
  European settlement, 82-83
  influence of rivers, 85
  urbanization, 84-85
  westward expansion, 84
maps, U.S. and world geography
  clothing, shelter, animals, 78
  geographic features, 79
  geographic terms, 79
  industrial areas of U.S., 75-76
  language of maps, 73-74
  location coordinates for U.S. capitals, 131-135
  location coordinates for world map, 128-130
  place names, 80, 113-121
  regions of U.S., 75, 160
  relief map of U.S., 79
  thematic world maps, 78-79
  transportation routes of U.S., 77-78
  world economics, 86
maps, world events
  connecting, 87
  contest, 89
  geography, 87

global grid quiz, 88
news sphere, 89
weekly headlines, 86
marathon map, 67
mileage chart, 68-69
multimedia kits, 170-171

neighborhood maps, 58-61

papier-mâché globes, 46
place names, 67-68, 113-121, 153-156
origins of, 79, 163-164
playground maps, 44-45
primary grades, basic concepts for,
7-8
print materials, 167-168

regions, U.S., 75, 159
relief maps, 45-46, 72-73
road maps, 39, 95-99
road signs, 15-16, 90-91, 98-102

safety symbols, 55-56, 150
scale and distance
calculating distance, 27-28, 29
different kinds of scale, 122-124
distance on a globe, 29-30
global grid, 33
inset maps, 28-29
longitude and global time, 33-35
planning a trip, 29
scale on a map, 27-28
understanding scale, 26-27,
122-124
using scale, 125
school maps, 51, 53
seating charts, 53-54
signs and symbols
highway map, 98-102
road signs, 93-94
safety, 55-56
weather maps, 40-41, 54-55,
144-147
See also symbols and legends
special-purpose maps, 37-38

state. See maps, state
states, spelling of, 103-106
supermarket floor plan, 62
symbols and legends, 13-19
flags, 16-17
globe, 13-14
photographs, drawings, maps,
14-15
road signs, 15-16, 93-95, 98-102
safety, 55-56
using, 17-19
weather, seasonal maps, 40-41, 54-
55, 144-147
See also signs and symbols

teachers, tips for, 11-12
terrain maps, 60. See also geography
thematic maps, 37-38
state, 69-70
world, 78-79
three-dimensional relief maps, 45-46
time zones, and longitude, 33-35
transportation routes, 77-78, 109-111,
139
travelog, 43, 152

United Nations, current members,
96-97
United States maps. See maps, U.S.
history; maps, U.S. and world
geography

weather maps, 40-41, 54-55, 144-147
barometric tendency, 145
Beaufort scale of winds, 146
weather report, 147
westward expansion, in U.S., 84
world
current events maps, 86-89
economic maps, 84
thematic maps, 78-79

zoo maps, 61